512.7
K639

Identification Numbers and Check Digit Schemes

Joseph Kirtland
Marist College

WITHDRAWI

Published and distributed by
The Mathematical Association of America

LIBRARY ST. MARY'S COLLEGE

CLASSROOM RESOURCE MATERIALS

This series provides supplementary material for students and their teachers—laboratory exercises, projects, historical information, textbooks with unusual approaches for presenting mathematical ideas, career information, and much more.

Committee on Publications
William Watkins, *Chair*

Classroom Resource Materials Editorial Board
Andrew Sterrett, Jr., *Editor*

Frank A. Farris	Stephen B Maurer
Julian Fleron	William A. Marion
Sheldon P. Gordon	Edward P. Merkes
Yvette C. Hester	Daniel E. Otero
Paul Knopp	Bruce P. Palka
Millianne Lehmann	Barbara J. Pence

101 Careers in Mathematics, edited by Andrew Sterrett

Archimedes: What Did He Do Besides Cry Eureka?, Sherman Stein

Calculus Mysteries and Thrillers, R. Grant Woods

Combinatorics: A Problem Oriented Approach, Daniel A. Marcus

A Course in Mathematical Modeling, Douglas Mooney and Randall Swift

Cryptological Mathematics, Robert Edward Lewand

Elementary Mathematical Models, Dan Kalman

Geometry From Africa: Mathematical and Educational Explorations, Paulus Gerdes

Identification Numbers and Check Digit Schemes, Joseph Kirtland

Interdisciplinary Lively Application Projects, edited by Chris Arney

Laboratory Experiences in Group Theory, Ellen Maycock Parker

Learn from the Masters, Frank Swetz, John Fauvel, Otto Bekken, Bengt Johansson, and Victor Katz

Mathematical Modeling in the Environment, Charles Hadlock

A Primer of Abstract Mathematics, Robert B. Ash

Proofs Without Words, Roger B. Nelsen

Proofs Without Words II, Roger B. Nelsen

A Radical Approach to Real Analysis, David M. Bressoud

She Does Math!, edited by Marla Parker

MW $37.70 7-23-01 (H)

MAA Service Center
P.O. Box 91112
Washington, DC 20090-1112
1-800-331-1MAA FAX: 1-301-206-9789

LIBRARY ST. MARY'S COLLEGE

Acknowledgement

Many people made this book possible, and I thank them all. First, I am extremely grateful to Pau-San Haruta of the Marist College English Department for her tireless efforts and valuable insights in the course of this project. Second, I thank Joe Gallian of the University of Minnesota–Duluth, whose numerous articles motivated this book and whose proofreading helped hone it. I also thank Louis Gross, Dorothy Buerk, and Kay Losey, whose constructive comments were greatly appreciated. Finally, I owe a great deal to April Sullivan, Marist College class of 1999, and Christina Sheedy, Marist College class of 2000, each of whom spent a summer helping me gather information. This book would not have been possible without them.

To my wife Cindy, son Timmy, and daughter Betsy.

Contents

Introduction **ix**

1 To the Student ix

2 To the Professor x

1 Identification Numbers and Check Digit Schemes **1**

1.1 Developing Identification Numbers 1

1.2 Types of Identification Numbers 2

1.3 Transmission Errors 4

1.4 Check Digits 5

2 Number Theory, Check Digit Schemes, and Cryptography **9**

2.1 Preliminaries 9

2.2 Integer Division 10

2.3 Modulo Arithmetic 19

2.4 US Postal Money Orders 26

2.5 Airline Ticket Identification Numbers 30

2.6 The Universal Product Code Check Digit Scheme 34

2.7 The ISBN Check Digit Scheme 41

2.8 Cryptography and the RSA Public-Key System 45

3 Functions, Permutations, and Their Applications **61**

3.1 Sets 61

3.2 Creating Identification Numbers 66

3.3 Functions 72

3.4 Permutations 81

3.5 The IBM Scheme 92
3.6 Graphs of Functions 98

4 Symmetry and Rigid Motions **105**
4.1 Symmetry 105
4.2 Symmetry and Rigid Motions 112

5 Group Theory and the Verhoeff Check Digit Scheme **121**
5.1 Fundamental Concepts 121
5.2 Cayley Tables 137
5.3 Powers and Orders of Group Elements 144
5.4 The Verhoeff Check Digit Scheme 152

6 Bibliography **167**

Index **173**

Introduction

0.1 To the Student

The ability to store, retrieve, and transmit data accurately is a central aspect of today's society. To make this process work efficiently, *identification numbers* are used to represent or encode information pertaining to products, documents, accounts, or individuals. One such system in place today generates a Universal Product Code (UPC) for every item sold in a grocery store. A UPC identifies not only the specific product but also the type of product and its manufacturer. This gives grocery stores a convenient and effective way to maintain inventory and keep track of sales (each store associates a price with the UPC that appears on the cash register when the product's bar code is scanned at the checkout counter). Numbers are also used to identify books (International Standard Book Numbers or ISBNs), individuals (social security numbers), library holdings, bank accounts, UPS packages, drivers' licenses, credit cards, and much more.

Given that identification numbers provide a convenient way to transmit information easily and accurately, they are recorded onto documents, typed or scanned into computers, sent via the Internet, or transmitted in some other fashion millions of times a day. Banks routinely transfer money electronically by using routing and account numbers, and consumers frequently complete sales with credit card numbers. Since these types of transactions occur so frequently, errors are bound to happen. For example, the number 12345 could be transmitted and incorrectly recorded as 12346 or as 12354. A bank would not want to transfer money into the wrong bank account, and consumers and retailers do not want charges billed to the wrong credit card account.

With our heavy reliance on identification numbers to transmit information and the likelihood that sooner or later a transmission error will occur, it is crucial to know when an identification number has been transmitted incorrectly. Specifically, the receiver of that number must have a way to determine whether the number received is incorrect. If no verification system has been established, the only way of knowing is to contact the sender. But contacting the sender is not always possible or feasible, and it may be time consuming. This difficulty has motivated the creation of methods that the receiver can use, independent of the sender, to recognize when an identification number has been transmitted incorrectly. The goal of this book is to present the mathematical methods, called *check digit schemes*, that do this.

Identification numbers, and the check digit schemes that detect when these numbers have been transmitted incorrectly, are crucial for the quick storage, retrieval, and transfer of vast amounts of information. In this book, a variety of check digit schemes are discussed. Check digit schemes vary in their ability to catch errors. Some, such as the airline ticket scheme, do not catch every occurrence of the most common type of error, while others, such as the ISBN scheme, catch most error patterns. Consequently, criteria for judging the reliability of check digit schemes are a central concern of this book.

0.2 To the Professor

This text is ideal for a liberal arts mathematics class. The book is organized to allow students to move from simple mathematical concepts and check digit schemes to more complex ideas. Not only are all mathematical concepts developed within the context of studying check digit schemes, but as each mathematical topic is studied, other applications are discussed. This will lead to a study of not only check digit schemes, but also "public key" cryptography systems, graphing data, presenting data, and symmetry.

Chapter 1 discusses a variety of identification number systems and establishes the mathematical terminology that will be used to study check digit schemes. In addition, the criteria used to determine the dependability of a scheme are presented.

Chapter 2 begins with a presentation of basic properties of integers and an introduction to modulo arithmetic. The concepts developed are then applied to an investigation of the check digit schemes used for United States' postal money orders, airline tickets, UPCs, and ISBNs. The reliability of each scheme is a central aspect of this discussion. Finding methods that address shortcomings of these schemes motivates the material covered in the remaining chapters. The same number-theoretic concepts are also applied to a discussion of cryptography, the art of sending secret messages. Special attention is paid to the RSA "public key" cryptography system, which is used to send sensitive data over the Internet.

In Chapter 3, sets, functions, and permutations are considered. These concepts play a role in the construction of more advanced check digit schemes and are central to *hashing functions.* A hashing function is the process used to take information and represent it as an identification number. The check digit scheme developed by IBM is also presented. In addition, the use of graphs to present and study functions and data is discussed.

The discussion in Chapter 4 is focused on symmetry and rigid motions. The notation established in Chapter 3 for permutations is used in a mathematical investigation of the symmetries of a variety of different shapes. The symmetries of a pentagon form the basis of the very reliable Verhoeff check digit scheme presented in Chapter 5. Furthermore, the use of rigid motions to create elaborate patterns will serve as an introduction to the discussion of group theory that begins Chapter 5.

In Chapter 5, an introduction to the fundamentals of group theory is presented. The concepts discussed, along with those presented in the previous chapters, culminate in the Verhoeff check digit scheme, the most sophisticated and reliable scheme considered in the book. The check digit scheme used with German money, which is based on the Verhoeff scheme, is also considered.

Along with the mathematical content described above, this book provides writing and group activities. These activities can be integrated into a student-centered approach. At the beginning of each section is a preliminary activity that has the students exploring and working with the concepts to be introduced. The notions that the students develop are then cultivated in that section. At the end of the section, traditional exercises, group activities, and writing assignments are given for further exploration.

Integral to this approach is the use of writing to develop and present mathematical understanding (see [17] and [24] for more information on the use of writing in the teaching of mathematics). Writing is not only a way to express mathematical understanding, but a way to develop that understanding. As each mathematical topic is investigated, the students use writing to improve and communicate their understanding of that topic and how it is applied. Students should write at the beginning, middle, and end of the learning process. The preliminary activities have them writing to investigate. This work is then rewritten or used to complete homework exercises and group activities, which are, in turn, used to complete larger writing assignments or essays. Through this writing and rewriting process, students gain a deeper understanding of mathematics and its diverse applications.

Writing to develop mathematical understanding will also improve communication skills. Many of the components of the mathematical process are rhetorical modes or critical writing strategies. Defining, serializing, classifying, comparing, generalizing, analyzing, and arguing are skills crucial to mathematics, but each is also a strategy that must be mastered to become an effective writer. Writing and paper assignments are included to help students develop each of these strategies.

1

Identification Numbers and Check Digit Schemes

1.1 Developing Identification Numbers

Preliminary Activity. Create a list of at least five different types of identifying numbers taken from a variety of different sources (e.g., product labels, magazines, and book covers). Bring these numbers to class. In groups, look at all the identification numbers collected, paying special attention to the lengths of each number, the types of characters used, and other defining characteristics. Write a short summary of your group's findings.

Identification numbers are used to identify individual items, specific products, people, accounts, or documents. Social security numbers, driver's license numbers, credit card numbers, International Standard Book Numbers (ISBNs), Universal Product Codes (UPCs), and bank account numbers are all examples of identification numbers. In addition, numbers are used to identify passports, United States (US) Post Office money orders, UPS packages, library books, merchandise, and many other items.

In today's "information age" with its high-powered computers, identification numbers are used to store and easily retrieve information and for many other purposes. They are used for easy identification of products, documents, or accounts and also for tracking and inventory purposes. For example, UPS numbers serve not only to identify specific packages but to track them as they travel to their destinations. Each package that a customer sends is assigned a number that can be easily scanned into and tracked by a computer. A central location can then provide information to customers so that they know when the package was picked up, where it is at any moment en route, and when the package is delivered. UPCs, found on most products, serve a variety of purposes. First of all, a UPC identifies the product, what type of product it is, and its manufacturer. Each store associates a price with the UPC, and that price appears on the cash register when the product's bar code is scanned at the checkout counter. Stores also use UPCs to maintain their inventory.

1.2 Types of Identification Numbers

An identification number usually takes the form of a string of digits, letters, symbols, or some combination of them. A digit is a number 0, 1, 2, 3, 4, 5, 6, 7, 8, or 9, and a symbol is a sign such as $*$, /, #, &, or $\backslash$. Together, digits, letters, and symbols form the *characters* that are used to create identification numbers. While its purpose is usually to provide a convenient and simple way to catalogue and identify items, an identification number can take many forms. It typically has either a *numeric* or an *alphanumeric* form:

- A numeric identification number can be a single positive number or a string of digits, sometimes broken up by spaces or dashes placed between the digits. Two numeric identification numbers are illustrated in Figures 1.1 and 1.2.
- An alphanumeric identification number has a string of digits, letters, and/or other symbols, as illustrated in Figures 1.3, 1.4, and 1.5.

In general, two pieces of information are needed in referring to an identification number:

- The *length*, or total number of characters.
- The *position* of each character.

The lowercase letter n denotes the *length* of an identification number, which is the total number of digits, letters, and symbols in the identification number.

FIGURE 1.1
Variety Pack of Jjust Jjuicy with UPC 0-53600-10054-0

FIGURE 1.2
US Postal Money Order with Identification Number 67021200988

JHMEC4521HS172003

FIGURE 1.3
The Vehicle Identification Number (VIN) for a 1987 Honda Civic

MOE**TH220DW

FIGURE 1.4
A Driver's License Number from the State of Washington

FIGURE 1.5
The ISBN 0-201-52032-X for David Lay's Linear Algebra Book

Example 1.2.1.

- The US postal money order number 67021200988 in Figure 1.2 has length $n = 11$.
- Both the UPC of 0-53600-10054-0 in Figure 1.1 and the Washington State driver's license number of MOE**TH220DW in Figure 1.4 have lengths $n = 12$.
- The Vehicle Identification Number (VIN) of JHMEC4521HS172003 in Figure 1.3 has length $n = 17$.

The *position* of each digit, letter, and symbol is also important. In general, if an identification number has a length of n, the number is denoted by $a_1a_2a_3 \ldots a_n$, where a_1 is the first character or digit in the number, a_2 is the second character or digit, a_3 is the third, and so on until a_n, which is the nth (last) character or digit in the number.

Example 1.2.2.

- US postal money order identification numbers always consist of 11 digits and are denoted $a_1a_2a_3a_4a_5a_6a_7a_8a_9a_{10}a_{11}$. For the number 67021200988,

$$a_1 = 6, \quad a_2 = 7, \quad a_3 = 0, \quad a_4 = 2, \quad a_5 = 1, \quad a_6 = 2,$$
$$a_7 = 0, \quad a_8 = 0, \quad a_9 = 9, \quad a_{10} = 8, \quad a_{11} = 8.$$

- Washington State driver's license numbers have length $n = 12$. We denote this by writing $a_1a_2a_3a_4a_5a_6a_7a_8a_9a_{10}a_{11}a_{12}$. For the number MOE**TH220DW,

$$a_1 = \text{M}, \quad a_2 = \text{O}, \quad a_3 = \text{E}, \quad a_4 = *, \quad a_5 = *, \quad a_6 = \text{T},$$
$$a_7 = \text{H}, \quad a_8 = 2, \quad a_9 = 2, \quad a_{10} = 0, \quad a_{11} = \text{D}, \quad a_{12} = \text{W}.$$

1.3 Transmission Errors

Every day, identification numbers are presented over the telephone, recorded onto documents, typed or scanned into computers, sent via the Internet, or transmitted in some other fashion. Each time this happens, there is a chance that one or more digits in the number will change or be rearranged as they move from one location to the other. For example, the UPC 0-53600-10054-0 on a product could be scanned into a computer as 0-53600-10059-0 or 0-53600-10045-0. In either case, an error in the number has occurred. In addition, less than honest individuals sometimes attempt to forge identification numbers (e.g., credit card numbers) for personal profit.

It is important to ensure that identification numbers are transmitted correctly and, in certain cases, not forged. If a UPC were scanned into a cash register incorrectly, the customer might be charged $10 for an item that should cost only $1.50. Banks, which transfer money electronically many times a day, do not want to transfer money into the wrong account. This type of error does happen. During the 1980s, Lt. Col. Oliver North, of Iran-Contra fame, gave US Assistant Secretary of State Elliot Abrams an incorrect Swiss bank account number for the purpose of depositing $10 million. North gave Abrams an account number that began "368" whereas the correct number began "386."

Transmission errors can occur when bar codes are scanned incorrectly, when numbers are written or typed in the wrong order, or when other mistakes are made. The most common types of errors that occur, based on a study presented in [26], are described below.

- A *single-digit error* occurs when one of digits in the number changes to a different value. 79.1% of all transmission errors that occur are single-digit errors.
- A *transposition-of-adjacent-digits error* occurs when two different side-by-side digits change places. 10.2% of all the transmission errors that occur are transposition-of-adjacent-digits errors.
- A *jump-transposition error* occurs when two different digits, separated by a third digit between them, change places. Of all the transmission errors that occur, 0.8% of them are jump-transposition errors.
- A *twin error* occurs when two identical side-by-side digits change to a different pair of identical digits. Of all transmission errors that occur, 0.5% of them are twin errors.
- A *phonetic error* occurs when two digits in the number, presented orally, are recorded incorrectly. For example, a person says "fourteen," but the recorder hears "forty." Of all the transmission errors that occur, 0.5% of them are phonetic errors.
- A *jump-twin error* occurs when two identical digits, separated by a third digit between them, change to a different pair of identical digits. Of all transmission errors that occur, 0.3% of them are jump-twin errors.

Table 1.1 gives an example of each type of error, using the six-digit identification number $a_1a_2a_3a_4a_5a_6 = 191433$. Table 1.2 lists all of these types of errors along with their relative frequencies, as based on a study presented in [26]. The lowercase letters a, b, and c represent single digits.

TABLE 1.1
Sample Errors

Error Type	Actual Number	Transmitted Error
Single digit	191433	191933
Transposition of adjacent digits	191433	191343
Jump transposition	191433	193413
Twin	191433	191455
Phonetic	191433	194033
Jump twin	191433	393433

TABLE 1.2
Common Error Patterns

Error Type	Form	Relative Frequency
Single digit	a → b	79.1%
Transposition of adjacent digits	ab → ba	10.2%
Jump transposition	abc → cba	0.8%
Twin	aa → bb	0.5%
Phonetic	a0 ↔ 1a*	0.5%
Jump twin	aca → bcb	0.3%

*For a = 2, . . . , 9.

Any one of the errors listed in Table 1.2 can occur when an identification number is transmitted. The receiver of that number has no way of knowing whether the number is correct unless the sender can be contacted. Contacting the sender is not always possible or feasible. This motivates the following goal: To develop methods by which the receiver can recognize when an identification number has been transmitted incorrectly. These methods, called *check digit schemes*, are not that difficult to develop. In fact, there are even established algorithms that identify errors and then correct them automatically. Information on "error correcting codes" can be found in [1] and [22].

1.4 Check Digits

Most check digit schemes append an extra digit or digits, called the *check digit(s)*, to the identification number and then use the digit(s) to check for errors after the number has been transmitted. To illustrate one possible method, consider the following scenario:

A company uses a three-digit number $a_1a_2a_3$ to identify each product it sells. To make sure these numbers are transmitted correctly, it adds a fourth digit (the check digit) a_4 to each product number, creating the four-digit identification number $a_1a_2a_3a_4$. To do this, the first three digits of the number are added,

$a_1 + a_2 + a_3$, and the check digit a_4 is assigned to be the last digit (the ones digit) of this sum.

For product number 854, $a_4 = 7$, since $8 + 5 + 4 = 17$ and 7 is the last digit in 17. The identification number is therefore 8547. The number 1090 would be a valid identification number, since the sum of the first three digits is 10 $(1+0+9 = 10)$ and 0, the last digit of 10, equals the last digit (the check digit) of 1090. The number 7352 would be invalid, however, since the sum of the first three digits is 15 $(7 + 3 + 5 = 15)$ and 5, the last digit of 15, does not equal 2, the last digit (the check digit) of 7352.

The check digit or digits can appear in any position in the number. However, the check often appears at the end as the last digit(s) in the identification number.

Example 1.4.1.

- For the US postal money order number of 67021200988 presented in Figure 1.2, the document number is 6702120098 and the last digit 8 is the check digit.
- For the UPC of 0-53600-10054-0 presented in Figure 1.1, the product number is 0-53600-10054 and the last digit 0 is the check digit.
- For all ISBNs, the check digit is either a digit or the letter X (the use of X will be explained in Chapter 2). For the ISBN presented in Figure 1.5, the book number is 0-201-52032 and the last digit X is the check digit.

These examples are all instances where the check digit is appended to the end of the number. However, check digits can appear anywhere in an identification number.

Example 1.4.2.

- For all Washington State driver's license numbers $a_1a_2a_3a_4a_5a_6a_7a_8a_9a_{10}a_{11}a_{12}$, the check digit is the tenth digit a_{10} of a 12-character alphanumeric identification number. For example, in the license number MOE**TH220DW presented in Figure 1.4, the tenth element $a_{10} = 0$ is the check digit.
- In a typical 17-digit VIN $a_1a_2a_3a_4a_5a_6a_7a_8a_9a_{10}a_{11}a_{12}a_{13}a_{14}a_{15}a_{16}a_{17}$, the check digit a_9 is in the ninth position. For the VIN JHMEC4521HS172003 presented in Figure 1.3, $a_9 = 1$ is the check digit.

Different identification number systems use different check digit schemes. Some are more effective than others. Given that single-digit errors and transposition-of-adjacent-digits errors account for almost 90% of all errors (see Table 1.2), any scheme that is developed should at the very least catch these two types of errors. Some of the simpler schemes, while better than nothing, do not even do this.

Of course, the more errors a scheme catches, the more complicated it will be. There are more advanced check digit schemes that catch all of the errors mentioned in Table 1.2. The main goal of this book is to develop the mathematics necessary to understand a variety of check digit schemes, examine their applications, and evaluate just how well they actually work.

Chapter 1 Exercises

1. Look around you. Give two examples, other than those presented in this chapter, of identification number systems used in everyday life. For each example, explain why it is needed and what problems could arise if the identification numbers were transmitted incorrectly.
2. Suppose you are an archivist at the Library of Congress and you have developed an identification number system to identify all of the different documents in the archives. Each document has a nine-digit $a_1a_2a_3a_4a_5a_6a_7a_8a_9$ identification number associated with it. The first seven digits $a_1a_2a_3a_4a_5a_6a_7$ identify the specific document, and the last two digits a_8 and a_9 are the check digits. For each document, the check digits a_8 and a_9 are determined by the sum of the digits in the document number. First calculate the sum $a_1 + a_2 + a_3 + a_4 + a_5 + a_6 + a_7$, and then assign a_8 to be the first (tens) digit and a_9 to be the second (ones) digit of this sum. In other words, the formula is $a_8a_9 = a_1 + a_2 + a_3 + a_4 + a_5 + a_6 + a_7$.

   For example, if the *Bill of Rights* has a document identification number of 2980162, then $a_8 = 2$ and $a_9 = 8$, since $2 + 9 + 8 + 0 + 1 + 6 + 2 = 28$. Consequently, the identification number associated with the *Bill of Rights* would be 298016228.

   If the sum of the first seven digits $(a_1+a_2+a_3+a_4+a_5+a_6+a_7)$ is a single digit a, then $a_8 = 0$ and $a_9 = a$. For example, consider the document number 1003201. Since the sum of the digits $1 + 0 + 0 + 3 + 2 + 0 + 1 = 7$, $a_8 = 0$, $a_9 = 7$, and the identification number would be 100320107.

   Consider the identification number 927361542. This would be an invalid number, as the sum of the first seven digits is $9 + 2 + 7 + 3 + 6 + 1 + 5 = 33$, which is not equal to the last two digits of 42. On the other hand, the identification number 927361533 would be a valid number.

   (a) Consider the identification numbers 102813321, 002315112, 109010920, and 398218085. Which of these are valid and which are invalid identification numbers? Be sure to show all your work and to explain your answer in the manner illustrated above.

   (b) Consider the document numbers 1092652, 0240170, and 2000003. Using the scheme described above, assign the check digits to each document number. Be sure to identify the check digits and then to write out the entire identification number including the check digits.

   (c) Will this scheme catch all single-digit errors? All transposition-of-adjacent-digits errors? Give examples of when this scheme will and will not catch an error.
3. You are in charge of all the computer equipment on the Marist College campus. To help keep track of all the equipment, you decide to develop an identification number system. To avoid transmission errors, this system will also incorporate a check digit scheme.

   Each computer on campus will have a five-digit $a_1a_2a_3a_4a_5$ computer number (CN) associated with it. The sixth digit a_6 will be the check digit and will be appended to the end of the CN to create a six-digit identification number $a_1a_2a_3a_4a_5a_6$.

Develop a check digit scheme for this number system. In other words, you must develop the process that creates the check digit a_6. There is no wrong answer as long as the method is mathematically sound. Give some examples of how it works. For example, if 27561 is a CN, what would the check digit be?

4. Most check digit schemes place the check digit as the last digit in the number. Is there a mathematical reason or other reason for this? Explain your answer.

Chapter 1 Paper Assignments

1. **Developmental Assignment: Defining.** Define the term *check digit scheme*. The audience for your definition is an undergraduate who is unfamiliar with the topics of identification numbers, check digits, and check digit schemes. Be sure to incorporate all the items necessary (identification number systems, check digits, and transmission errors) for understanding check digit schemes. Provide examples at each stage of the definition.
2. **Summarizing.** Write a short essay that will explain to someone not familiar with check digit schemes the motivation behind the creation of check digit schemes. Clearly specify the relationship between check digit schemes and identification numbers.
3. **Developmental Assignment: Argumentation.** Find an identification number system that, as far as you know, does not contain a check digit scheme. Make a case for the creation of a new identification number system that incorporates a check digit scheme. That is, give reasons why such a system/scheme might be useful and for whom.

Chapter 1 Group Activity

1. Each group will generate three examples, other than those presented in this chapter, of identification number systems used today. For each example, explain why it is needed and what problems could arise if the numbers were transmitted incorrectly.

Further Reading

Vinzant, C., What Hidden Meanings Are Embedded in Your Social Security Number?, *Fortune*, 139, 1999, 32.

2

Number Theory, Check Digit Schemes, and Cryptography

This chapter begins with an investigation of integer division and modulo arithmetic. We then explore check digit schemes that employ the number theoretic concepts we have developed. The chapter ends with an application of these concepts to cryptography.

2.1 Preliminaries

In mathematics, many different types of numbers can be used. Different situations and settings require different types of numbers. When the number of items in a store display is being counted, the numbers $0, 1, 2, 3$, and so on are used. However, to present the batting averages of major league ballplayers, we need fractions and their decimal equivalents. For example, if a player has 136 hits in 418 at bats, his average is $\frac{136}{418} = .325$.

There are basically six different sets or types of numbers. Only five will be mentioned here. The brace symbols $\{$ and $\}$ will be used to indicate a set or collection of numbers. The left brace $\{$ indicates the beginning of the set and the right brace $\}$ indicates the end. The ellipsis symbol $\ldots$ indicates that the pattern of numbers established continues the same way without end or until a specified ending value is reached.

- **Counting or Natural Numbers:** These are the numbers that are used for counting. They are $\{1, 2, 3, 4, \ldots\}$.
- **Whole Numbers:** This is the set of natural numbers including 0. We write this set as $\{0, 1, 2, 3, 4, \ldots\}$.
- **Integers:** This is the set of whole numbers and their negatives. They are $\{\ldots, -4, -3, -2, -1, 0, 1, 2, 3, 4, \ldots\}$.
- **Rational Numbers:** This is the set containing those numbers that can be expressed

as a fraction p/q, where p and q are integers with $q \neq 0$. Some examples of rational numbers are $\frac{1}{2}$, $\frac{11}{106}$, $\frac{-2}{5}$, and $\frac{4}{1} = 4$.

- **Real Numbers** When a fraction is expressed as a decimal, it either terminates or repeats. For example, the fraction $\frac{3}{4} = 0.75$ terminates and the fraction $\frac{8}{11} = 0.727272\ldots$ repeats (denoted $0.\overline{72}$). Many numbers in the number line do not repeat and do not terminate; $\sqrt{2} = 1.4142135\ldots$, $\sqrt{24} = 4.8989794\ldots$, and $\pi = 3.1415926\ldots$ are some examples. Numbers whose decimal notation does not terminate and does not repeat are called *irrational numbers. Real numbers* are the set of all rational numbers combined with the set of all irrational numbers.

Most of the mathematics in this book involves integers. Counting numbers, which are integers greater than 0, are often referred to as the *positive integers*. Whole numbers, which are integers greater than or equal to 0, are sometimes referred to as *nonnegative integers*.

Recall the four operations from arithmetic: addition, subtraction, multiplication, and division. These operations, along with integers and whole numbers, will form the basis of our investigation into check digit schemes.

Notation. To perform an arithmetic operation when one or more of the numbers in the expression is not specifically known, we denote these numbers by using the lowercase letters $x, y, z, a, b, \ldots$, called *variables*. For example, to indicate the addition of 5 to another number, whose value is unknown, $x + 5$ is written, where x denotes the unknown number. It could just as well be written as $y + 5$, where y denotes the unknown number.

Exercise 2.1

1. Explain why every integer is a rational number.

Further Reading

Burton, D. M., *The History of Mathematics,* WCB McGraw-Hill, Boston, 1999.

2.2 Integer Division

Preliminary Activity.

1. Each member of a group chooses one of these four lists of integers to investigate and answers the following questions.

   List 1. $2, 8, 18, 22, 100, 330, 502$

   List 2. $3, 9, 12, 18, 33, 150, 330$

   List 3. $5, 15, 45, 100, 215, 330, 575$

   List 4. $11, 33, 77, 132, 330, 891$

(a) What is the common characteristic of all of the integers in your list? (HINT: Think in terms of division.)

(b) What differentiates the first integer in your list from the others (besides the fact that it is the smallest)? How is it different?

(c) There is a number in your list that is divisible only by the first number in your list and the first number in one of the other lists. What number is it? How can this number be obtained by using the other two?

2. The group members now compare notes. What is the common thread in each of your answers to the questions from part (1)?

3. The number 330 is in all four lists. What properties does it have that cause it to be in each list? Express 330 as a product of "smaller" integers.

As mentioned in the previous section, given two numbers (12 and 3, for example), they can both be added together ($12 + 3 = 15$), one can be subtracted from the other ($12 - 3 = 9$), one can be divided by the other ($12 \div 3 = 4$), or they can be multiplied together ($12 \cdot 3 = 36$). For the moment, division involving two integers will be explored.

Consider $12 \div 3$, which equals 4. Another way to express this is $12/3 = 4$. Both symbols $\div$ and $/$ are used to denote division. Now, why is $12/3 = 4$? One way to look at it is to write $12/3$ as a fraction and note that $12 = 4 \cdot 3$. Since $3/3 = 1$, $12/3 = 4$:

$$12/3 = \frac{12}{3} = \frac{4 \cdot 3}{3} = \frac{4 \cdot \not{3}}{\not{3}} = 4 \cdot 1 = 4.$$

Another way to determine that $12/3 = 4$ is by using long division. Clearly, 3 goes into 12 a total of 4 times with a remainder of 0:

$$\begin{array}{r} 4\ r = 0 \\ 3\overline{)12} \\ \underline{-12} \\ 0 \end{array}$$

Since the remainder when 12 is divided by 3 is 0, $12/3 = 4$. It is also said that 3 *goes into* 12 *evenly* or that 3 *divides* 12.

In this case, dividing the integer 12 by the integer 3 (that is, $12/3$) yields the integer 4. However, division of one integer by another does not always result in an integer. When the integer 123 is divided by the integer 5 ($123/5$), the result is not an integer ($123/5 = 24.6$). Another way to see that the result is not an integer is to use long division:

$$\begin{array}{r} 24\ r = 3 \\ 5\overline{)123} \\ \underline{-10} \\ 23 \\ \underline{-20} \\ 3 \end{array}$$

Thus, when 123 is divided by 5, the result is not an integer as the remainder is not 0.

These two cases, a remainder of 0 and a remainder of 3 (an integer not equal to 0), need to be differentiated. But first, another observation can be made. In the first case, 3 divided evenly into 12. It was also noted that $12 = 4 \cdot 3$, or that 12 is a *multiple* of 3. In the second case, this does not happen. The integer 123 is *not a multiple* of 5. If 123 were to be a multiple of 5, there would be some integer n such that $123 = n \cdot 5$. Given that $24 \cdot 5 = 120$, that $25 \cdot 5 = 125$, and that 123 is an integer between 120 and 125, there is no integer n such that $n \cdot 5$ equals 123. Thus 123 is not a multiple of 5. Since 12 is a multiple of 3, 3 is said to *divide* 12. Since 123 is not a multiple of 5, 5 *does not divide* 123.

Definition 2.2.1. *An integer y **divides** an integer x if there is another integer n such that $x = n \cdot y$. This is denoted by $y \mid x$. If y does not divide x, it is denoted $y \nmid x$.*

Example 2.2.2.

- $2 \mid 34$ or 2 *divides* 34, since $34 = 17 \cdot 2$.
- $7 \mid 315$ or 7 *divides* 315, since $315 = 45 \cdot 7$.
- $25 \mid -150$ or 25 *divides* -150, since $-150 = -6 \cdot 25$.
- $1 \mid 23$ or 1 *divides* 23, since $23 = 23 \cdot 1$.
- $23 \mid 23$ or 23 *divides* 23, since $23 = 1 \cdot 23$.
- $23 \mid -23$ or 23 *divides* -23, since $-23 = -1 \cdot 23$.

Example 2.2.3.

- $2 \nmid 35$ or 2 *does not divide* 35, since $34 = 17 \cdot 2$ and $36 = 18 \cdot 2$.
- $7 \nmid 318$ or 7 *does not divide* 318, since $315 = 45 \cdot 7$ and $322 = 46 \cdot 7$.
- $25 \nmid -155$ or 25 *does not divide* -155, since $-150 = -6 \cdot 25$ and $-175 = -7 \cdot 25$.
- $5 \nmid 3$ or 5 *does not divide* 3, since $0 = 0 \cdot 5$ and $5 = 1 \cdot 5$.

In terms of integer division, some special integers need to be mentioned. They are defined next.

Definition 2.2.4. *An integer is **prime** if it is greater than* 1 *and if the only positive integers that divide it are* 1 *and itself. An integer greater than* 1 *that is not prime is **composite**.*

There are an infinite number of prime numbers. Examples include $2, 3, 5, 7, 11, 13, 17, 23, 127, 239$, and 457. Integers such as 8 ($8 = 2 \cdot 4$) and 33 ($33 = 3 \cdot 11$) are not prime. Consequently, 8 and 33 are composite numbers.

Determining which integers are prime can be a real challenge. The Greek mathematician Eratosthenes of Cyrene (276–194 B.C.) developed a technique for finding all the prime numbers less than or equal to a certain whole number n. He is famous not only for this technique, called the *Sieve of Eratosthenes*, but also for being director of the library at Alexandria and for his use of Euclidean geometry to come up with a fairly accurate measure of the circumference of the earth.

The *Sieve* is used by first listing all the numbers from 2 to n ($2, 3, 4, 5, 6, \ldots, n$). Once this is done, we put a box around 2, the first prime number in the list. Since 2 multiplied by an integer greater than or equal to 2 ($2 \cdot 2 = 4$, $3 \cdot 2 = 6$, $4 \cdot 2 = 8$, etc.) will not be prime, these numbers can be crossed off the list. The next number after 2 that is not crossed off is 3. We box it, as it is prime. Then we cross off all the multiples of 3. We continue in this manner until the largest integer less than or equal to $n/2$ is reached.[1] Then we box all the numbers that have not been crossed off. All the boxed numbers will be prime.

To illustrate how this process works, all the prime numbers less than or equal to 29 will be found. Since $29/2 = 14.5$, 14 is the largest integer less than or equal to $29/2$, and this process will be followed until the integer 14 is reached. First, all the integers between 2 and 29 are listed:

2 3 4 5 6 7 8 9 10 11 12 13 14 15
16 17 18 19 20 21 22 23 24 25 26 27 28 29

Box 2 and cross off all multiples of 2:

$\boxed{2}$ 3 ~~4~~ 5 ~~6~~ 7 ~~8~~ 9 ~~10~~ 11 ~~12~~ 13 ~~14~~ 15
~~16~~ 17 ~~18~~ 19 ~~20~~ 21 ~~22~~ 23 ~~24~~ 25 ~~26~~ 27 ~~28~~ 29

The next number after 2 that is not a multiple of 2 (not crossed off the list) is 3. Consequently, it is prime. Box it and cross off all the multiples of 3. Some of them were crossed off the list earlier:

$\boxed{2}$ $\boxed{3}$ ~~4~~ 5 ~~6~~ 7 ~~8~~ ~~9~~ ~~10~~ 11 ~~12~~ 13 ~~14~~ ~~15~~
~~16~~ 17 ~~18~~ 19 ~~20~~ ~~21~~ ~~22~~ 23 ~~24~~ 25 ~~26~~ ~~27~~ ~~28~~ 29

The next number after 3 that is not a multiple of 2 or 3 (not crossed off the list) is 5. Consequently, it is prime. Box it and cross off all the multiples of 5. Some of them were crossed off the list earlier:

$\boxed{2}$ $\boxed{3}$ ~~4~~ $\boxed{5}$ ~~6~~ 7 ~~8~~ ~~9~~ ~~10~~ 11 ~~12~~ 13 ~~14~~ ~~15~~
~~16~~ 17 ~~18~~ 19 ~~20~~ ~~21~~ ~~22~~ 23 ~~24~~ ~~25~~ ~~26~~ ~~27~~ ~~28~~ 29

The next number after 5 that is not a multiple of 2, 3, or 5 (not crossed off the list) is 7. Consequently, it is prime. Box it and cross off all the multiples of 7. In this case, all of them have been already crossed off the list:

$\boxed{2}$ $\boxed{3}$ ~~4~~ $\boxed{5}$ ~~6~~ $\boxed{7}$ ~~8~~ ~~9~~ ~~10~~ 11 ~~12~~ 13 ~~14~~ ~~15~~
~~16~~ 17 ~~18~~ 19 ~~20~~ ~~21~~ ~~22~~ 23 ~~24~~ ~~25~~ ~~26~~ ~~27~~ ~~28~~ 29

The next number after 7 that is not a multiple of 2, 3, 5, or 7 (not crossed off the list) is 11. Consequently, it is prime. Box it and cross off all the multiples of 11. In this case, all of them have been already crossed off the list:

$\boxed{2}$ $\boxed{3}$ ~~4~~ $\boxed{5}$ ~~6~~ $\boxed{7}$ ~~8~~ ~~9~~ ~~10~~ $\boxed{11}$ ~~12~~ 13 ~~14~~ ~~15~~
~~16~~ 17 ~~18~~ 19 ~~20~~ ~~21~~ ~~22~~ 23 ~~24~~ ~~25~~ ~~26~~ ~~27~~ ~~28~~ 29

---

[1] Let $m > n/2$, then the first multiple of m is $2 \cdot m$. Since $m > n/2$, this results in $2 \cdot m > 2 \cdot n/2 = n$. Since $2 \cdot m > n$, it is not a number less than n and need not be considered.

The next number after 11 that is not a multiple of 2, 3, 5, 7, or 11 (not crossed off the list) is 13. Consequently, it is prime. Box it and cross off all the multiples of 13. In this case, all of them have been already crossed off the list:

$$\begin{array}{cccccccccccccc}
\boxed{2} & \boxed{3} & \cancel{4} & \boxed{5} & \cancel{6} & \boxed{7} & \cancel{8} & \cancel{9} & \cancel{10} & \boxed{11} & \cancel{12} & \boxed{13} & \cancel{14} & \cancel{15} \\
\cancel{16} & 17 & \cancel{18} & 19 & \cancel{20} & \cancel{21} & \cancel{22} & 23 & \cancel{24} & \cancel{25} & \cancel{26} & \cancel{27} & \cancel{28} & 29
\end{array}$$

Since 14 is crossed off the list, box all the numbers that remain on the list. They are all the prime numbers less than or equal to 29:

$$\begin{array}{cccccccccccccc}
\boxed{2} & \boxed{3} & \cancel{4} & \boxed{5} & \cancel{6} & \boxed{7} & \cancel{8} & \cancel{9} & \cancel{10} & \boxed{11} & \cancel{12} & \boxed{13} & \cancel{14} & \cancel{15} \\
\cancel{16} & \boxed{17} & \cancel{18} & \boxed{19} & \cancel{20} & \cancel{21} & \cancel{22} & \boxed{23} & \cancel{24} & \cancel{25} & \cancel{26} & \cancel{27} & \cancel{28} & \boxed{29}
\end{array}$$

There is a fairly simple, yet sometimes time-consuming, technique to determine whether a single positive integer m is prime. Suppose m is composite. Then some integer $b > 1$ divides m ($b \mid m$). Thus there is a positive integer a such that $m = ab$. Suppose that $a > \sqrt{m}$ and $b > \sqrt{m}$, then $a \cdot b > \sqrt{m} \cdot \sqrt{m} = m$. This would be a contradiction, as $m = ab$. Thus either $a \le \sqrt{m}$ or $b \le \sqrt{m}$. In other words, if m is composite, there will be a positive integer, in fact a prime number, less than or equal to $\sqrt{m}$ that divides m. Consequently, to determine whether a single positive integer m is prime, show that no prime number less than or equal to $\sqrt{m}$ divides m. The following example illustrates this technique.

Example 2.2.5.

- 101: The primes less than or equal to $\sqrt{101} = 10.0498\ldots$ are 2, 3, 5, and 7. It is a simple process to show that none of them divides 101. Thus, 101 is prime.
- 221: The primes less than or equal to $\sqrt{221} = 14.86606\ldots$ are 2, 3, 5, 7, 11, and 13. It is easy to show that 13 divides 221 ($221 = 17 \cdot 13$). Thus 221 is not prime.

Hints.

- The digits in a number add up to 3 if and only if the number is divisible by 3. Using this fact, 141 is divisible by 3, as $1 + 4 + 1 = 6 = 2 \cdot 3$, and 142 is not divisible by 3, as $1 + 4 + 2 = 7$, a number not divisible by 3.
- A number ending in 5, other than the number 5 itself, can never be prime.
- A number ending in 0, 2, 4, 6, or 8, other than the number 2 itself, can never be prime.

As shall be seen, prime numbers are used in a variety of ways. One simple property shared by all positive integers n is that each can be written as a product of prime numbers. For instance, take the number 60: 60 is not prime and is thus divisible by a prime number. In particular, the prime number 2 divides 60 or $60 = 2 \cdot 30$. Now consider 30, which also is not prime. The prime number 2 divides 30 or $30 = 2 \cdot 15$. Since $60 = 2 \cdot 30$ and $30 = 2 \cdot 15$, $60 = 2 \cdot 2 \cdot 15$. Likewise, 15 is not prime. It is divisible by 3 and can be written $15 = 3 \cdot 5$. Thus $60 = 2 \cdot 30 = 2 \cdot 2 \cdot 15 = 2 \cdot 2 \cdot 3 \cdot 5 = 2^2 \cdot 3 \cdot 5$. As a result, 60 is now expressed as a product involving the primes 2, 3, and 5. The expression $2^2 \cdot 3 \cdot 5$ is called the *prime factorization* of 60:

$$\begin{aligned} 60 &= 60 \\ &= 2 \cdot 30 \\ &= 2 \cdot 2 \cdot 15 \\ &= 2 \cdot 2 \cdot 3 \cdot 5 \\ &= 2^2 \cdot 3 \cdot 5. \end{aligned}$$

Proposition 2.2.6. *Every integer $n > 1$ can be expressed as a unique product of prime numbers $n = p_1^{r_1} p_2^{r_2} \cdots p_t^{r_t}$, where $p_1, p_2, \ldots, p_t$ are the distinct primes dividing n with $p_1 < p_2 < \cdots < p_t$, and r_i, for $1 \leq i \leq t$, is the power of each prime p_i. The product $p_1^{r_1} p_2^{r_2} \cdots p_t^{r_t}$ is called the* ***prime factorization*** *of n.*

Sketch of Proof. If n is prime, it is its own prime factorization. If n is composite, then there is a smallest prime number p that divides n. Since $p \mid n$, there is a number m_1 such that $m = p \cdot m_1$. If m_1 is prime, $p \cdot m_1$ is the prime factorization. If m_1 is not prime, there is a smallest prime number q that divides it. Since $q \mid m_1$, there is an integer m_2 such that $m_1 = q \cdot m_2$. Thus $m = p \cdot m_1 = p \cdot q \cdot m_2$. This process is continued until m is written as a product of primes. □

Example 2.2.7.

- The prime factorization of 7 is 7 (7 is itself a prime number).
- The prime factorization of 182 is $2 \cdot 7 \cdot 13$.
- The prime factorization of 360 is $2^3 \cdot 3^2 \cdot 5$.
- The prime factorization of 81 is 3^4.

Up to this point, fundamental concepts involving division and the integers have been investigated. The integer y divides the integer x or $y \mid x$, if $x = n \cdot y$ for some integer n. For example, $4 \mid 12$ as $12 = 3 \cdot 4$, but 5 does not divide 12 as there is no multiple of 5 that results in 12. Positive integers greater than 1 that can only be divided by 1 and themselves are called prime. And every number, whether prime or composite (not prime), can be written as a product of primes. These concepts are central as our investigation into the properties of the integer continues.

Definition 2.2.8. *Two integers x and y are* ***relatively prime*** *if there is no positive integer greater than 1 that divides them both.*

Example 2.2.9.

- 6 and 35 are relatively prime. The only integers greater than 1 that divide 6 are 2, 3, and 6 ($6 = 2 \cdot 3 = 1 \cdot 6$), and the only integers greater than 1 that divide 35 are 5, 7, and 35 ($35 = 5 \cdot 7 = 1 \cdot 35$.)
- 6 and 14 are not relatively prime as $2 \mid 6$ and $2 \mid 14$.

The prime factorizations of the positive integers x and y are useful in determining whether x and y are relatively prime. For Example 2.2.9, in the second case, it was determined that 6 and 14 are not relatively prime. Given their prime factorizations, $6 = 2{\cdot}3$ and $14 = 2 \cdot 7$, it can readily be seen that 2 divides both numbers and thus that 6 and 14 are not relatively prime. In the first case presented in Example 2.2.9, it was determined that 6 and 35 are relatively prime. Given their prime factorizations, $6 = 2{\cdot}3$ and $35 = 5{\cdot}7$, it can readily be seen that no positive integer greater than 1 divides both of them and thus that they are relatively prime.

Example 2.2.10.

- 90 and 735 are not relatively prime. Consider each integer's prime factorization:

$$90 = 2 \cdot 3^2 \cdot 5 \qquad \text{and} \qquad 735 = 3 \cdot 5 \cdot 7^2.$$

  Each integer's prime factorization indicates that 3 divides both integers. In fact, since the integers 2 and 3 both divide 90 and 735, $2 \cdot 3 = 6$ also divides both 90 and 735.

- 550 and 273 are relatively prime. Consider each integer's prime factorization:

$$550 = 2 \cdot 5^2 \cdot 11 \qquad \text{and} \qquad 273 = 3 \cdot 7 \cdot 13.$$

  Each integer's prime factorization indicates that no prime number, and consequently no integer greater than 1, divides both of them.

Recall the work done earlier in this chapter on division. Since $60 = 12 \cdot 5$, $5 \mid 60$. It is also easy to show that $13 \nmid 60$:

$$\begin{array}{r} 4 \; r=8 \\ 13\overline{)60} \\ \underline{-52} \\ 8 \end{array}$$

Another way to say this is that 13 goes into 60 a total of 4 times with a remainder of 8. This is written $60 = 4 \cdot 13 + 8$. In fact, this can be done with any two integers as long as division by 0 is not attempted.

Theorem 2.2.11 (Division Algorithm). *For any two integers x and y, where $y > 0$, there exist unique integers q and r such that*

$$x = q \cdot y + r \qquad \text{where } 0 \le r \le y - 1.$$

The number q is called the *quotient* and the number r is called the *remainder when x is divided by y* or just called the *remainder*. Note that the remainder can be 0.

Example 2.2.12.

- $x = 60$ and $y = 5$: $60 = 12 \cdot 5 + 0$.
  Here the quotient $q = 12$ and the remainder $r = 0$.
- $x = 60$ and $y = 13$: $60 = 4 \cdot 13 + 8$.
  Here the quotient $q = 4$ and the remainder $r = 8$.

- $x = 7$ and $y = 10$: $7 = 0 \cdot 10 + 7$.
  Here the quotient $q = 0$ and the remainder $r = 7$.
- $x = 227$ and $y = 12$: $227 = 18 \cdot 12 + 11$.
  Here the quotient $q = 18$ and the remainder $r = 11$.
- $x = -87$ and $y = 6$: $-87 = -15 \cdot 6 + 3$.
  Here the quotient $q = -15$ and the remainder $r = 3$.

Exercises 2.2

1. Find all the prime numbers between 2 and 200 and give a brief explanation of how they were found.
2. Which of the following integers are prime? Give a short explanation of how each answer was obtained.

   (a) 151 (b) 201 (c) 207 (d) 211 (e) 305
3. Write a short paragraph explaining why a positive integer ending in a 5, except 5 itself, cannot be prime.
4. Write a short paragraph explaining why a positive integer ending in a 0, 2, 4, 6, or 8, except 2 itself, cannot be prime.
5. Determine whether the following statements are true or false. Be sure to indicate how you got your answer. Simply saying "true" or "false" is not enough.

   (a) $9 \mid 99$ (b) $6 \mid 172$ (c) $4 \mid 29$ (d) $24 \mid 192$

   (e) $10 \mid 105$ (f) $3 \nmid 21$ (g) $13 \nmid 55$ (h) $3 \nmid 22$
6. Find the quotient and remainder when

   (a) 138 is divided by 15; (b) 35 is divided by 5;

   (c) 100 is divided by 17; (d) -38 is divided by 9.
7. (a) Find two examples of a pair of integers a and b with the properties $a \mid b$ and $b \mid a$.

   (b) Generalize the findings from part (a) of this problem. In other words, given two integers a and b where $a \mid b$ and $b \mid a$, what property must a and b satisfy?
8. Consider two integers n and m such that $n = p_1p_2$ and $m = q_1q_2$, where p_1, p_2, q_1, and q_2 are distinct primes. For example, n could be $6 = 2 \cdot 3$ ($p_1 = 2$ and $p_2 = 3$), and m could equal $35 = 5 \cdot 7$ ($q_1 = 5$ and $q_2 = 7$). Show that n and m must always be relatively prime.
9. Suppose the prime factorization of an integer n contains the prime factors 2, 3, and 11. In other words, $n = 2^r \cdot 3^s \cdot 11^t$ for positive integers $r \geq 1, s \geq 1$, and $t \geq 1$. Show that

   (a) 6 divides n; (b) 22 divides n;

   (c) 7 does not divide n; (d) 35 does not divide n.

Paper Assignments 2.2

1. **Summarizing.** Write a summary of Sections 1 and 2 from this chapter. Among other terms, your summary should include the following, but not necessarily in this order: *natural numbers, whole numbers, integers, rational numbers, real numbers, prime numbers, relatively prime, divides, quotient,* and *remainder*. The explanation of these concepts should not appear as a list but should be connected, such that your summary takes the form of a brief essay. For the general framework of this essay, use the rationale given at the beginning of this chapter as to why we need to know these concepts. Where notation is used, follow the format from the text.

2. **Serializing.** The Division Algorithm states that for any two integers x and y, where $y > 0$, there exist unique integers q and r such that $x = q \cdot y + r$, where $0 \leq r \leq y - 1$. Write an essay that explains to someone unfamiliar with the Division Algorithm the steps to follow in order to find the quotient and remainder when an integer x is divided by an integer y ($y > 0$). This can be done in two stages. First, present a few cases where the quotient and remainder are found for specific pairs of integers (for example, show the steps that need to be followed to obtain the quotient and remainder when 157 is divided by 25). As these examples are presented, comment on the steps used to find the remainder and the quotient. In the second stage of the essay, exhibit the steps that must be followed in order to find the quotient and remainder when an arbitrary integer x is divided by an integer y. This can be done by presenting the Division Algorithm with the variables x, y, q, and r and connecting them with the specific values used in the first stage of the essay.

3. **Analysis.** The Division Algorithm states that for any two integers x and y, where $y > 0$, there exist unique integers q and r such that $x = q \cdot y + r$, where $0 \leq r \leq y - 1$. Write an essay that explains why the quotient q and the remainder r are unique. This can be done by taking two specific values for x and y, say $x = 157$ and $y = 25$, and showing why the quotient and the remainder obtained when 157 is divided by 25 are unique.

4. **Argumentation.** Consider the integers 3917, 90091, and 10101. For each integer, create an argument as to why the number is or is not prime. This will involve stating whether the number is prime or not and then presenting the work done (evidence) that supports your answer. Terms and techniques introduced in this chapter should be an integral part of the essay.

Group Activities 2.2

1. Consider three integers a, b, and c that satisfy the property that $a \mid b$ and $b \mid c$.

   (a) Come up with three examples of integers a, b, and c that satisfy the conditions that $a \mid b$ and $b \mid c$. The integers $a = 2$, $b = 6$, and $c = 12$ are one such example, as $a \mid b$ or $2 \mid 6$ and $b \mid c$ or $6 \mid 12$. In each case, find the integers l_1 and l_2 such that $b = l_1 \cdot a$ and $c = l_2 \cdot b$.

(b) Show for each set of numbers found in part (a) that $a \mid c$. In each case, find the integer l such that $c = l \cdot a$.

(c) Given the work done above, make a conjecture as to what property a and c will satisfy, given that $a \mid b$ and $b \mid c$. Look at all three examples generated so far.

(d) Write a few sentences that support the conjecture made in part (c).

2. Consider a prime integer p and two other integers a and b such that $p \mid ab$ (ab is the product of a and b).

(a) Come up with three examples of a prime integer p and integers a and b that satisfy the condition that $p \mid ab$. In each case, find the integer t where $ab = t \cdot p$. For example, given $p = 5$, $a = 3$, and $b = 10$, $p \mid ab$ or $5 \mid 30$ and $30 = 6 \cdot 5$ ($t = 6$).

(b) In terms of the concepts introduced so far, is there a relationship between p and a or p and b in each of the examples created? Make a conjecture as to what conclusion can be made when a prime integer p divides ab, where a and b are integers. Support and test this conjecture by creating a few more examples.

(c) Generalize the work done in parts (a) and (b) and construct an argument that supports the conjecture formed in part (b).

Further Reading

Roberts, J., *Lure of the Integers*, The Mathematical Association of America, Washington, D.C., 1992.

2.3 Modulo Arithmetic

Preliminary Activity.

1. In a group, each member chooses one of these four lists of integers to investigate and answers the following questions.

List 1. $5, 21, 37, 85, 125$

List 2. $1, 13, 37, 61, 121$

List 3. $8, 26, 44, 53, 98$

List 4. $4, 11, 32, 67, 81$

(a) What is the common characteristic of all of the integers in your list? For List 1, think in terms of division by 8. For List 2, think in terms of division by 12. For List 3, think in terms of division by 9. For List 4, think in terms of division by 7.

(b) What differentiates the first integer in your list from the others (besides the fact that it is the smallest)?

2. The group members now compare notes. What is the common thread to each of your answers to the questions from part (1)?

3. Determine a simple equation or method that could be used to generate each integer on each list.

Suppose that today is Friday, August 5, and someone asks the question "What day of the week will it be in seven days?" This is easy to answer since a week consists of seven days. Thus in seven days it will be Friday. Likewise, Friday would also be the answer to the question "What day of the week will it be in 21 days?" Again the answer is Friday because a week has seven days and 21 days represent exactly three weeks. For the question "What day of the week will it be in eight days?" the answer is Saturday by similar reasoning.

To answer the day-of-the-week question posed above, the number 7 was used to represent the unit 1. This was done because 7 or seven days represent one week. By the Division Algorithm, $8 = 1 \cdot 7 + 1$. In other words, since the number 8 is one more than the number 7, eight days from now will be Saturday (one day after Friday). Mathematically this is the case because 8 has the remainder of 1 when divided by 7.

This idea of letting a number other than 1 represent a unit is used in many places. For example, there are 24 hours in 1 day, 12 inches in 1 foot, and so on. Many mathematical calculations do the same. A formal investigation of these ideas is presented here. When given an integer x, we are interested in what the remainder r will be when x is divided by a positive integer n.

The integer 37 has a remainder of 7 when divided by 10 ($37 = 3 \cdot 10 + 7$). The integer 48 has a remainder of 3 when divided by 5 ($48 = 9 \cdot 5 + 3$). But writing "37 has a remainder of 7 when divided by 10" and "48 has a remainder of 3 when divided by 5" is fairly cumbersome. Here is a better way.

Notation. Let x and n be integers with $n > 0$. The remainder r obtained when x is divided by n is denoted by $x \pmod{n}$.

Using this modulo notation, we can write $37 \pmod{10} = 7$ and $48 \pmod{5} = 3$. Below are some other examples.

Example 2.3.1.

- $79 \pmod{6} = 1$: $79 = 13 \cdot 6 + 1$.
- $132 \pmod{15} = 12$: $132 = 8 \cdot 15 + 12$.
- $6 \pmod{20} = 6$: $6 = 0 \cdot 20 + 6$.

In many of the applications examined in this book, mod 10 arithmetic is central.

Example 2.3.2.

- $79 \pmod{10} = 9$: $79 = 7 \cdot 10 + 9$.
- $132 \pmod{10} = 2$: $132 = 13 \cdot 10 + 2$.
- $7 \pmod{10} = 7$: $7 = 0 \cdot 10 + 7$.

Two numbers can also be added, subtracted, or multiplied modulo n. $(a+b) \pmod n$, $(a-b) \pmod n$, or $(a \cdot b) \pmod n$ result in the remainder when $a+b$, $a-b$, or $a \cdot b$ is divided by n. In fact,

$$\begin{aligned}(a+b) \pmod n &= \big((a \pmod n) + (b \pmod n)\big) \pmod n\\ (a-b) \pmod n &= \big((a \pmod n) - (b \pmod n)\big) \pmod n\\ (a \cdot b) \pmod n &= \big((a \pmod n) \cdot (b \pmod n)\big) \pmod n.\end{aligned}$$

Given these two slightly different methods, the one used depends on which approach will make the calculations the easiest. For example, $(22+15) \pmod{10}$ will result in the remainder when $22+15$ is divided by 10. Since $22+15 = 37 = 3 \cdot 10 + 7$, $(22+15) \pmod{10} = 7$. This could also be calculated in the following manner:

$$\begin{aligned}(22+15) \pmod{10} &= \big((22 \pmod{10}) + (15 \pmod{10})\big) \pmod{10}\\ &= (2+5) \pmod{10}\\ &= 7 \pmod{10}\\ &= 7.\end{aligned}$$

Similarly, $(22 \cdot 15) \pmod{10} = 330 \pmod{10} = 0$ can also be calculated as follows:

$$\begin{aligned}(22 \cdot 15) \pmod{10} &= \big((22 \pmod{10}) \cdot (15 \pmod{10})\big) \pmod{10}\\ &= (2 \cdot 5) \pmod{10}\\ &= 10 \pmod{10}\\ &= 0.\end{aligned}$$

Now consider the calculation $8^4 \pmod{10}$. The answer can be determined by computing 8^4 ($8^4 = 4096$) and then finding the remainder when 8^4 is divided by 10. That is, compute $8^4 \pmod{10} = 4096 \pmod{10} = 6$. This computation of $8^4 \pmod{10}$ can be simplified, using the fact that the power 4 is itself a power of 2 ($4 = 2^2$). As a result, $8^4 = 8^2 \cdot 8^2$ and $\pmod{10}$ can be applied after each multiplication. In other words, since $8^4 = 8^2 \cdot 8^2$, we can write

$$\begin{aligned}8^4 \pmod{10} &= 8^2 \cdot 8^2 \pmod{10}\\ &= \big((8^2 \pmod{10}) \cdot (8^2 \pmod{10})\big) \pmod{10}\\ &= \big((64 \pmod{10}) \cdot (64 \pmod{10})\big) \pmod{10}\\ &= \big((4 \pmod{10}) \cdot (4 \pmod{10})\big) \pmod{10}\\ &= (4 \cdot 4) \pmod{10}\\ &= 16 \pmod{10}\\ &= 6.\end{aligned}$$

Now consider the calculation of $49^{11} \pmod{85}$. It can be simplified,[2] as presented below, using the fact that $49^{11} = 49^8 \cdot 49^2 \cdot 49$:

$$\begin{aligned}
49 \pmod{85} &= 49;\\
49^2 \pmod{85} &= 2401 \pmod{85} = 21;\\
49^4 \pmod{85} &= 49^2 \cdot 49^2 \pmod{85} = 21 \cdot 21 \pmod{85}\\
&= 441 \pmod{85} = 16;\\
49^8 \pmod{85} &= 49^4 \cdot 49^4 \pmod{85} = 16 \cdot 16 \pmod{85} = 1.
\end{aligned}$$

Since $49^{11} \pmod{85} = (49^8 \cdot 49^2 \cdot 49) \pmod{85}$,

$$\begin{aligned}
49^{11} \pmod{85} &= \Big((49^8 \pmod{85}) \cdot (49^2 \pmod{85}) \cdot (49 \pmod{85})\Big) \pmod{85}\\
&= (1 \cdot 21 \cdot 49) \pmod{85}\\
&= 1029 \pmod{85}\\
&= 9.
\end{aligned}$$

Using these techniques to find $8^4 \pmod{10}$ and $49^{11} \pmod{85}$ simplifies much of the arithmetic involved. In addition, as the next example indicates, the "moding out" can be also be done before the numbers are multiplied together.

Example 2.3.3.

- Compute $27^5 \pmod{15}$.

$$\begin{aligned}
27^5 \pmod{15} &= \big(27 \pmod{15}\big)^5 \pmod{15}\\
&= 12^5 \pmod{15}\\
&= \Big((12^2 \pmod{15}) \cdot (12^2 \pmod{15}) \cdot 12 \pmod{15}\Big) \pmod{15}\\
&= \Big((144 \pmod{15}) \cdot (144 \pmod{15}) \cdot 12 \pmod{15}\Big) \pmod{15}\\
&= (9 \cdot 9 \cdot 12) \pmod{15}\\
&= 972 \pmod{15}\\
&= 12.
\end{aligned}$$

Now consider the numbers 37, 85, and 117:

$$\begin{aligned}
37 &= 3 \cdot 10 + 7,\\
85 &= 8 \cdot 10 + 5,\\
117 &= 11 \cdot 10 + 7.
\end{aligned}$$

In this case, $37 \pmod{10} = 117 \pmod{10}$ (both 37 and 117 have the same remainder of $r = 7$ when divided by 10). This is written $37 \equiv 117 \pmod{10}$, since writing $\pmod{10}$ twice is redundant. In contrast, $37 \pmod{10} \neq 85 \pmod{10}$ or $37 \not\equiv 85 \pmod{10}$ as 37 and 85 have different remainders when divided by 10.

---

[2] To determine $49^2 \pmod{85}$ with a calculator, enter 49×49 to obtain 2401, then divide 2401 by 85 to obtain 28.247058. Finally, enter $2401 - (28 \times 85)$ to obtain 21

If two integers x_1 and x_2 have the same remainder when divided by an integer n, they are called *congruent.*

Remark. The equals sign $=$ and the equivalence sign $\equiv$ are different. The equals sign, $=$, is used when two numbers or expressions are identical. For example, $x + 3 = 5$ or $10 = 10$. The equivalence sign, $\equiv$, is used when two expressions or numbers share a common property. Here the property is that the two integers have the same remainder when divided by a third positive integer.

Definition 2.3.4. *Let n be a positive integer. Given two integers x_1 and x_2, x_1* ***is congruent to x_2 modulo*** *n, written $x_1 \equiv x_2 \pmod{n}$, if x_1 and x_2 both have the same remainder when divided by n.*

When x_1 and x_2 are not congruent modulo n, $x_1 \not\equiv x_2 \pmod{n}$ is written.

Example 2.3.5.

- $23 \equiv 39 \pmod{4}$: $23 = 5 \cdot 4 + 3$ and $39 = 9 \cdot 4 + 3$.
- $19 \equiv 82 \pmod{7}$: $19 = 2 \cdot 7 + 5$ and $82 = 11 \cdot 7 + 5$.
- $161 \equiv 1,331 \pmod{15}$: $161 = 10 \cdot 15 + 11$ and $1,331 = 88 \cdot 15 + 11$.
- $3 \equiv 23 \pmod{10}$: $3 = 0 \cdot 10 + 3$ and $23 = 2 \cdot 10 + 3$.
- $23 \not\equiv 41 \pmod{4}$: $23 = 5 \cdot 4 + 3$ and $41 = 10 \cdot 4 + 1$ (different remainders).
- $23 \not\equiv 30 \pmod{10}$: $23 = 2 \cdot 10 + 3$ and $30 = 3 \cdot 10 + 0$ (different remainders).

Exercises 2.3

1. Calculate each of the following expressions.

   (a) $15 \pmod{10}$ (b) $23 \pmod{7}$ (c) $106 \pmod{13}$
   (d) $54 \pmod{9}$ (e) $(25 - 9) \pmod{7}$ (f) $(13 + 39) \pmod{10}$
   (g) $(12 \cdot 18) \pmod{6}$ (h) $13^4 \pmod{9}$ (i) $33^7 \pmod{10}$

2. Determine whether the following statements are true or false. Be sure to indicate how you got the answer you did. Simply saying "true" or "false" is not enough.

   (a) $17 \equiv 77 \pmod{10}$ (b) $5 \equiv 17 \pmod{4}$
   (c) $138 \equiv 225 \pmod{3}$ (d) $100 \equiv 301 \pmod{25}$

3. Describe all of the solutions a to the equation $a \equiv 6 \pmod{10}$, where a is an arbitrary integer. Listing all the solutions is not enough. Write a short paragraph that indicates how you got your answer.

4. (a) Calculate each of the following expressions.

   $$20 \pmod{10} \qquad 34 \pmod{10} \qquad 89 \pmod{10}$$

(b) Based on the work done in part (a), how many possible remainders are there when an integer is divided by 10? Part (a) gives only three possible remainders.

(c) Using your answer from part (b), how many possible solutions x are there to the equation $a \pmod{10} = x$, where a is an arbitrary integer?

5. (a) Calculate each of the following expressions.

$$42 \pmod{21} \qquad 69 \pmod{21} \qquad 80 \pmod{21} \qquad 104 \pmod{21}$$

(b) Based on the work done in part (a), how many possible remainders are there when an integer is divided by 21? Part (a) gives only four possible remainders.

(c) Using your answer from part (b), how many possible solutions x are there to the equation $a \pmod{21} = x$, where a is an arbitrary integer?

Paper Assignments 2.3

1. **Summarizing.** Write a short summary of Section 3 from this chapter. Among other terms, your summary should include the following, but not necessarily in this order: *remainder, congruence*, and *modulo*. The explanation of these concepts should not appear as a list but should be connected, such that your summary takes the form of a brief essay. Where notation is used, follow the format from the text.
2. **Developmental Assignment.** Combine the summary of sections 1 and 2 of this chapter with the summary written for section 3. The concepts and terms introduced in sections 1 and 2 are needed in section 3. All of these concepts will be central to the study of cryptography and check digit schemes.
3. **Description and Argumentation.** Besides the days of the week, can you think of other sets of numbers used around us whose manipulation can usefully be described in terms of modulo arithmetic? Give examples of this application of modulo arithmetic and explain why it would be beneficial in the situations that you have selected.

Group Activities 2.3

1. (a) Consider the integers 38 and 71. It is easy to show that $71 \equiv 38 \pmod{11}$ since $38 = 3 \cdot 11 + 5$ and $71 = 6 \cdot 11 + 5$. Note that $71 - 38 = 33 = 3 \cdot 11$ or that $11 \mid (71 - 38)$.

(b) Will what happened in part (a) always occur when two integers x_1 and x_2 are congruent modulo n? In other words, if integers x_1 and x_2 satisfy the equation $x_1 \equiv x_2 \pmod{n}$, will n always divide $x_1 - x_2$? Test this hypothesis. With each group member using different values for n, come up with three sets of integers x_1 and x_2, such that $x_1 \equiv x_2 \pmod{n}$. Then for each set of integers, find $x_1 - x_2$ and determine whether n divides $x_1 - x_2$.

(c) Using your results from part (b), write down the property that $x_1 - x_2$ must satisfy when $x_1 \equiv x_2 \pmod{n}$.

(d) Write down the calculations that support the statement from part (c). (HINT: When $x_1 \equiv x_2 \pmod{n}$, $x_1 = q_1 \cdot n + r$ and $x_2 = q_2 \cdot n + r$ for integers q_1, q_2, and r. Then write $x_1 - x_2$, replacing x_1 with $q_1 \cdot n + r$ and x_2 with $q_2 \cdot n + r$, and simplify the expression.)

2. (a) Find the values of x that satisfy the following equation: $23 \equiv x \pmod{4}$.

   i. By definition, the solutions x must have the same remainder as 23 has when divided by 4. The first step is to find the remainder when 23 is divided by 4.

   ii. Based on your work in part (i), use trial and error to find at least two solutions x to the equation $23 \equiv x \pmod{4}$.

   iii. Show that $x = -5, -1, 3, 7, 11, 15$, and 19 are also solutions.

   iv. What is the common bond among all the solutions mentioned in part (iii)? Use the Division Algorithm to write each solution x as $x = q \cdot 4 + r$.

   v. Using the results from part (iv), derive a formula or method for finding all the solutions x to the equation $23 \equiv x \pmod{4}$.

   (b) Find the values of x that satisfy the following equation: $57 \equiv x \pmod{10}$.

   i. By definition, the solutions x must have the same remainder as 57 has when divided by 10. The first step is to find the remainder when 57 is divided by 10.

   ii. Based on your work in part (i), use trial and error to find at least two solutions x to the equation $57 \equiv x \pmod{10}$.

   iii. Show that $x = -3, 7, 17, 27, 37, 47$, and 57 are also solutions.

   iv. Comparing the solutions presented in (iii) to each other, what is the common bond among them? Use the Division Algorithm to write each solution x as $x = q \cdot 10 + r$.

   v. Using the results from part (iv), derive a formula or method for finding all the solutions x to the equation $57 \equiv x \pmod{10}$.

   (c) Find all the solutions x to the equation $105 \equiv x \pmod{31}$.

Further Reading

Ogilvy, C. S., and Anderson, J. T., *Excursions in Number Theory,* Dover, New York, 1966.

Ore, O., *Invitation to Number Theory,* Random House, New York, 1967.

Comment. In the first three sections of this chapter, fundamental concepts involving division and the integers were investigated. We began with a discussion of integer division, prime integers, and prime factorizations. These concepts were central as our investigation into the properties of the integers continued. Studying the Division Algorithm led to the concept that two integers x and y are equivalent modulo n (written $x \equiv y \pmod{n}$) if x and y have the same remainder r when each is divided by n.

The next four sections of this chapter will study the application of these number theoretic ideas to four different check digit schemes. While the first two schemes (US postal money orders and airline ticket numbers) are fairly simple, they do not fare well with regards to catching all of the types of errors listed in Table 1.2. The last two schemes (UPC and ISBN) are more sophisticated and do a better job in catching errors.

2.4 US Postal Money Orders

Preliminary Activity. Recall Exercise 2 at the end of Chapter 1 where a check digit scheme for the Library of Congress was presented. In that exercise, each document had a nine-digit $a_1a_2a_3a_4a_5a_6a_7a_8a_9$ identification number associated with it. The first seven digits $a_1a_2a_3a_4a_5a_6a_7$ identified the specific document and the last two digits a_8 and a_9 were the check digits. For each document, the check digits a_8 and a_9 were determined by the sum of the digits in the document number. First the sum $a_1+a_2+a_3+a_4+a_5+a_6+a_7$ was calculated, and then a_8 was assigned to be the first (tens) digit and a_9 was assigned to be the second (ones) digit of this sum. In other words, the check digits are calculated as $a_8a_9 = a_1 + a_2 + a_3 + a_4 + a_5 + a_6 + a_7$. For example, if the Bill of Rights had a document identification number of 2980162, then $a_8 = 2$ and $a_9 = 8$ since $2+9+8+0+1+6+2 = 28$. Consequently, the identification number associated with the Bill of Rights would be 298016228. Now consider the following situation:

> The Library of Congress wants to shorten, by one digit, the length of its identification numbers. There is now to be an eight-digit number $a_1a_2a_3a_4a_5a_6a_7a_8$ associated with each document. Since the Library of Congress still wants $a_1a_2a_3a_4a_5a_6a_7$ to identify the specific document, now there will be only one check digit a_8 instead of two. Develop a check digit scheme for this new system that involves, as the previous method did, the sum $a_1+a_2+a_3+a_4+a_5+a_6+a_7$ of the first seven digits of the number. Keep in mind the number theoretic techniques introduced at the beginning of this chapter as you develop a method that will determine the check digit a_8 from the sum $a_1+a_2+a_3+a_4+a_5+a_6+a_7$.

As mentioned in Chapter 1, the US Post Office uses an identification number system for postal money orders. The identification number is 11 digits long. The first ten digits are the document number, and the last digit is the check digit. Recall the money order described in Chapter 1. Its 11-digit identification number is 67021200988. The first ten digits 6702120098 identify the document and the last digit 8 is the check digit.

The US Post Office uses a "mod 9" check digit scheme [3], [7]. The check digit is the remainder when the sum of digits in the document number is divided by 9.

Definition 2.4.1. *Let $a_1a_2a_3a_4a_5a_6a_7a_8a_9a_{10}$ be the ten-digit document number associated with a US postal money order. The check digit a_{11}, which is appended to this number to create the 11-digit identification number $a_1a_2a_3a_4a_5a_6a_7a_8a_9a_{10}a_{11}$, is determined by*

$$a_{11} = (a_1 + a_2 + a_3 + a_4 + a_5 + a_6 + a_7 + a_8 + a_9 + a_{10}) \pmod 9.$$

Consider the identification number 67021200988 which describes a document numbered $a_1a_2a_3a_4a_5a_6a_7a_8a_9a_{10} = 6702120098$. The check digit $a_{11} = 8$, since

$$\begin{aligned}(a_1 + a_2 + a_3 + a_4 + a_5 + a_6 + a_7 + a_8 + a_9 + a_{10}) \pmod 9 \\ &= (6+7+0+2+1+2+0+0+9+8) \pmod 9 \\ &= 35 \pmod 9 = 8\end{aligned}$$

or $35 = 3 \cdot 9 + 8$.

Consider another possible money order number, 31059112852. If this is to be a valid number, the remainder when $3+1+0+5+9+1+1+2+8+5$ is divided by 9 must be equal to the check digit of 2. However,

$$3+1+0+5+9+1+1+2+8+5 = 35 = 3 \cdot 9 + 8$$

or $35 \pmod 9 = 8$. Since the remainder is 8, which does not equal the check digit of 2, this is an invalid number.

This method is also used to generate the check digit. Suppose that the ten-digit number to be used to identify a money order is 2791400953. The complete identification number will be $2791400953C$, where C is the check digit. C must satisfy the property

$$(2+7+9+1+4+0+0+9+5+3) \pmod 9 = C.$$

Thus to find C simply find the remainder when $2+7+9+1+4+0+0+9+5+3$ is divided by 9. Since

$$2+7+9+1+4+0+0+9+5+3 = 40 = 4 \cdot 9 + 4,$$

$C = 4$. The complete identification number is 27914009534.

The question now is: How good is this scheme?

Recall that the goal for any check digit scheme is to catch all the errors listed in Table 1.2. At the very least, a scheme should catch all the single-digit and transposition-of-adjacent-digits errors. Consider the valid money order identification number that was just created. Listed below are two different single-digit errors.

Correct Number:	27**9**14009534	279140095**5**34
	↓	↓
Incorrect Number:	27**0**14009534	279140098**8**34

In the first case, the money order scheme does not catch the error, as

$$(2+7+0+1+4+0+0+9+5+3) \pmod 9 = 31 \pmod 9 = 4$$

$(31 = 3 \cdot 9 + 4)$, which is the check digit.

In the second case, the error is caught, as

$$(2+7+9+1+4+0+0+9+8+3) \pmod 9 = 43 \pmod 9 = 7$$

$(43 = 4 \cdot 9 + 7)$ and 7 does not match the check digit of 4.

Thus it does not catch all single-digit errors. However, it does come very close. This scheme will catch all single-digit errors except those where a 9, in any position except the check digit position, is replaced by a 0, or vice versa.

So, why does this scheme not catch the types of single-digit errors mentioned above? Consider the following case, where a 9, in any position except the check digit position, is replaced by a 0 in an arbitrary US postal money order identification number.

Correct Number:	$a_1 \ldots \underline{\mathbf{9}} \ldots a_{10}a_{11}$
	↓
Incorrect Number:	$a_1 \ldots \underline{\mathbf{0}} \ldots a_{10}a_{11}$

By definition, the correct number $a_1 \ldots 9 \ldots a_{10}a_{11}$ will satisfy the condition $a_{11} = (a_1 + \cdots + 9 + \cdots + a_{10}) \pmod 9$. Let $n = a_1 + \cdots + 9 + \cdots + a_{10}$. Thus n has the remainder of a_{11} when divided by 9 or $n = q \cdot 9 + a_{11}$ for some quotient q.

To see if the single-digit error in the incorrect number $a_1 \ldots 0 \ldots a_{10}a_{11}$ is caught, the sum $m = a_1 + \cdots + 0 + \cdots + a_{10}$ is computed and tested by determining whether $a_{11} = (a_1 + \cdots + 0 + \cdots + a_{10}) \pmod 9$ or $a_{11} = m \pmod 9$. The sum of digits in n is almost the same as the sum of digits in m. The only difference is that the number 9 in the sum denoted by n is replaced by a 0 in the sum denoted by m. As a result, m is 9 less than n, or $m = n - 9$. Since $n = q \cdot 9 + a_{11}$,

$$\begin{aligned} m &= n - 9 \\ &= q \cdot 9 + a_{11} - 9 \\ &= q \cdot 9 - 9 + a_{11} \\ &= (q - 1) \cdot 9 + a_{11}. \end{aligned}$$

Thus, $a_{11} = m \pmod 9$ and the computation indicates that the incorrect number is valid. So this type of single-digit error is not caught. The case where 0 is replaced by 9 is similar.

This money order scheme does a much poorer job at detecting transposition-of-adjacent-digits errors. It catches these types of errors only when the transposition involves the check digit. Consider the following examples.

Correct Number:	27**91**4009534	279140095**34**
	↓	↓
Incorrect Number:	27**19**4009534	279140095**43**

In the first case, this scheme does not catch the error, as

$$(2 + 7 + 1 + 9 + 4 + 0 + 0 + 9 + 5 + 3) \pmod 9 = 40 \pmod 9 = 4$$

$(40 = 4 \cdot 9 + 4)$. In the second case, the error is caught, as

$$(2 + 7 + 9 + 1 + 4 + 0 + 0 + 9 + 5 + 4) \pmod 9 = 41 \pmod 9 = 5$$

$(41 = 4 \cdot 9 + 5)$, which is not the listed check digit of 3.

Exercises 2.4

1. The following two numbers are claimed to be US postal money order identification numbers. Using the US postal money order check digit scheme, determine which is a valid number and which is invalid.

   67021200112   10962801047

2. The number 3980062110 is to be used to identify a US postal money order. Using the US postal money check digit scheme, assign a check digit to this identification number. Be sure to identify the check digit and then to write out the entire document number.
3. List all the numbers that could be check digits using the US postal money order check digit scheme. Explain how you got your answer.
4. Explain why the US postal money order check digit scheme catches only transposition-of-adjacent-digits errors that involve the check digit. (HINT: Try a few examples and look at the sum of the digits.)

Group Activities 2.4

1. At the end of this section, it was mentioned that the US postal money order check digit scheme catches transposition-of-adjacent-digits errors only when they involve the check digit. The goal of this activity is to determine why.

   (a) Consider the following US postal money order identification numbers.

   57037409875 10521477898 01513268245 90123212125

   (b) Each member of the group chooses one of the numbers listed and performs the following tasks.

      i. Do the calculation to show that your number is a valid identification number.

      ii. Rewrite the number twice: Show a transposition-of-adjacent-digits error that does not involve the check digit, and show one that does.

      iii. Apply the check digit scheme to each of the incorrect numbers generated in step (ii). The error involving the check digit will be caught, while the other one will not be caught. Be careful to write out each calculation.

      iv. Compare the calculation for the invalid numbers to the calculation for the valid one. What similarities do you see? What differences do you see?

   (c) As a group, form a conjecture as to why the US postal money order check digit scheme catches transposition-of-adjacent-digits errors only when they involve the check digit.

Further Reading

Gallian, J. A., Check Digit Methods, *International Journal of Applied Engineering Education*, 5(4), 1989, 503–505.

Gallian, J. A., and Winters, S., Modular Arithmetic in the Marketplace, *American Mathematical Monthly*, 95, 1988, 548–551.

2.5 Airline Ticket Identification Numbers

Airline tickets have 15-digit identification numbers associated with them. The first 14 digits identify the ticket and the 15th (the last one) is the check digit.

In Figure 2.1, the identification number and check digit are 0-001-1300696719-4. The first digit, here a 0, is the coupon number. A coupon number of 1 identifies a ticket for the first flight of the trip, a 2 identifies a ticket for the second flight of the trip, and so on. The coupon number 0 identifies the customer receipt. The second part of the identification number, here 001, identifies the airline. The third part, here 1300696719, is the document number. And the last digit, here 4, is the check digit. Airline tickets use a "mod 7" check digit scheme [3], [7]. The same scheme is used by Federal Express and UPS.

Definition 2.5.1. *Let* $a_1\text{-}a_2a_3a_4\text{-}a_5a_6a_7a_8a_9a_{10}a_{11}a_{12}a_{13}a_{14}$ *be the 14-digit number associated with an airline ticket. The check digit* a_{15}*, which is appended to this number to create the identification number* $a_1\text{-}a_2a_3a_4\text{-}a_5a_6a_7a_8a_9a_{10}a_{11}a_{12}a_{13}a_{14}\text{-}a_{15}$*, is determined by*[3]

$$a_{15} = a_1a_2a_3a_4a_5a_6a_7a_8a_9a_{10}a_{11}a_{12}a_{13}a_{14} \pmod 7.$$

For the airline ticket number 0-001-1300696719-4 from Figure 2.1, the number

$$a_1a_2a_3a_4a_5a_6a_7a_8a_9a_{10}a_{11}a_{12}a_{13}a_{14} = 00011300696719,$$

and the check digit $a_{15} = 4$, since

$$00011300696719 = 11300696719 = 1614385245 \cdot 7 + 4.$$

Consider the 15-digit number 1-012-359102991-0. If this is to be a valid airline ticket number, the remainder when 1012359102991 is divided by 7 must be 0. However,

$$1012359102991 = 144622728998 \cdot 7 + 5,$$

or $1012359102991 \pmod 7 = 5$. Since the remainder is 5, not 0, the number is invalid.

This method is also used to generate the check digit. Suppose that the 14-digit number to identify an airline ticket is 0-004-2871911233. The complete document number will be 0-004-2871911233-C, where C is the check digit. C must satisfy the property that 00042871911233 or $42871911233 \pmod 7) = C$. To find C, simply find the remainder when 42871911233 is divided by 7. Since

$$42871911233 = 6124558747 \cdot 7 + 4,$$

$C = 4$. The complete document number is 0-004-2871911233-4.

The question now is: How good is this scheme?

The goal is to catch all the errors mentioned in Table 1.2. If not all, at the very least a scheme should catch all the single-digit errors and transposition-of-adjacent-digits errors.

---

[3] The airline identification number is not always included in the check digit calculation. The only way to tell is to calculate the check digit with and without the airline identification number included.

PASSENGER TICKET AND BAGGAGE CHECK
SUBJECT TO CONDITIONS OF CONTRACT
NOT TRANSFERABLE
PASSENGER RECEIPT
ARC FLIGHT COUPON
ISSUED BY AMERICAN AIRLINES
TOUR CODE XX OK XX
AGENT CODE A33583723
NAME OF ISSUING AGENT CARD TVL AGCY
PLACE OF ISSUE POUGHKEEPSIE NY
ISO CODE US DATE OF ISSUE 12NOV92
NAME OF PASSENGER KIRTLAND/JOSEPH
PNR/CARRIER CODE SNFXG5/AA
FARE BASIS/TICKET DESIGNATOR VE14NR
FCI SERV CARR. ID L 0011/
X/O FROM CARRIER FLIGHT CLASS DATE TIME STATUS NOT VALID BEFORE NOT VALID AFTER
NOT VALID FOR THIS IS YOUR RECEIPT
X/O TO
**TRANSPORTATION*
ISSUING AGENT ID B1ND*73
ENDORSEMENTS/RESTRICTIONS
NO REFUND/CHG SUBJ USD 25 FEE PLUS FARE DIFF
FP IK5329041908049411*0993/ 000034 /FCSWF AA X/CH
I AA SAT190.91 AA X/CHI AA SWF190.91VE14NR 381.82 E
ND
FARE USD 381.82
TAX 38.18
TAX
TOTAL USD 420.00
EQUIV FARE PD.
ALLOW PCS WT UNCKD
STOCK CONTROL NO. TX 889 CK
25519657225
CPN DOCUMENT NUMBER CK
0 001 1300696719 4
0020039 AKG
BOARDING PASS
NAME OF PASSENGER KIRTLAND/JOSEPH
SWF
FROM
XORD AA1801 V 12JANVE14NR
OSAT AA1493 V 12JANVE14NR
TO
XORD AA426 V 17JANVE14NR
SWF AA1064 V 17JANVE14NR
CARRIER
CARRIER FLIGHT CLASS DATE TIME
GATE SEAT SMOKE
PCS WT UNCKD BAGGAGE ID NUMBER
NOT VALID FOR TRAVEL
0 001 1300696719 4
AA33583723
IT IS UNLAWFUL TO PURCHASE OR RESELL THIS TICKET FROM/TO ANY ENTITY OTHER THAN THE ISSUING CARRIER OR ITS AUTHORIZED AGENTS

FIGURE 2.1
Airline Ticket with Identification Number 0-001-1300696719-4

Consider the valid airline ticket number 0-004-2871911233-4 that was just created. Here are are two different single-digit errors.

Correct Number:	0-004-2871<u>**9**</u>11233-4	0-004-2<u>**8**</u>71911233-4
	↓	↓
Incorrect Number:	0-004-2871<u>**5**</u>11233-4	0-004-2<u>**1**</u>71911233-4

In the first case, this scheme does catch the error, because $42871511233 \pmod 7 = 5$ (that is, $42871511233 = 6124501604 \cdot 7 + 5$), and 5 does not equal the check digit of 4. In the second case, the error is not caught, as $42171911233 \pmod 7 = 4$ (that is, $42171911233 = 6024558747 \cdot 7 + 4$).

Even though it does not catch all single-digit errors, the ticket scheme does come close. Let a be a single digit in an airline ticket number that is replaced by b to create an invalid number (a single-digit error where a is replaced by b). This scheme will catch all single-digit errors except those where $\mid a - b \mid = 7$. In the first example, where the single-digit error was caught, $a = 9$, $b = 5$, and

$$\mid a - b \mid = \mid 9 - 5 \mid = \mid 4 \mid = 4.$$

In the second example, where the single-digit error was not caught, $a = 8$, $b = 1$, and

$$\mid a - b \mid = \mid 8 - 1 \mid = \mid 7 \mid = 7.$$

This airline ticket scheme does a similar job at detecting transposition-of-adjacent-digits errors. It will catch most transposition errors. Let a and b be two adjacent digits in an airline ticket number. An error involving the transposition of these two adjacent digits (that is, $\ldots ab \ldots \rightarrow \ldots ba \ldots$) will be caught except when $\mid a - b \mid = 7$.

Correct Number:	0-004-287<u>**0**</u>911233-3 → 0-004-28<u>**70**</u>911233-3	0-004-287091<u>**12**</u>33-3
	↓	↓
Incorrect Number:	0-004-28<u>**07**</u>911233-3	0-004-287091<u>**21**</u>33-3

In the first case, this scheme does not catch the error, as $42807911233 \pmod 7 = 3$ (that is, $42807911233 = 6115415890 \cdot 7 + 3$), and 3 is the listed check digit. Here $a = 7$, $b = 0$, and

$$| a - b | = | 7 - 0 | = | 7 | = 7.$$

In the second case, however, the error is caught, as $42870912133 \pmod 7 = 0$ (that is, $42870912133 = 6124416019 \cdot 7 + 0$). In this case, $a = 1$, $b = 2$, and

$$| a - b | = | 1 - 2 | = | -1 | = 1 \neq 7.$$

Exercises 2.5

1. Suppose that the following two numbers are airline ticket identification numbers. Using the airline ticket check digit scheme, determine which is a valid number and which is an invalid number.

   0-037-1222494405-0 0-037-1222495505-0

2. The number 0-037-1222494315 is used to identify an airplane ticket. Use the airline ticket check digit scheme to assign a check digit to this number. Be sure to identify the check digit and then to write out the entire document number.
3. List all the possible numbers that could be check digits for the airline check digit scheme. Explain how you got your answer.

Paper Assignment 2.5

1. **Comparison.** You are employed by an organization that wants to adopt a check digit scheme. Your job is to write a report called a feasibility study, in which you recommend to the company either the US postal order scheme or the airline ticket scheme. To do this, compare the US postal order check digit scheme with the airline ticket check digit scheme, paying attention to the following factors:

   (a) ease of use (the length of, time involved with, and complexity of the calculations);

   (b) strength (in terms of the types of errors it can detect).

   In making your evaluation, you might want to consider the type of organization that will use your scheme.

Group Activity 2.5

1. At the end of this section, it was mentioned that the airline ticket check digit scheme does not catch all single-digit errors. When a is a single digit in an airline ticket number that is replaced by b to create an invalid number (a single-digit error where

a is replaced by b), the scheme will not catch this error when $| a - b |= 7$. The goal of this activity is to understand why.

Since airline ticket numbers are large, the exercise will be completed with much smaller four-digit numbers, $a_1a_2a_3a_4$. The scheme works in the same manner. The check digit a_4 is appended to the three-digit number $a_1a_2a_3$ to create the identification number $a_1a_2a_3a_4$ such that a_4 satisfies the equation $a_4 = a_1a_2a_3 \pmod 7$. These numbers are easier to work with, but the mathematical concept is the same.

Consider these four valid identification numbers, along with their associated single-digit errors (Incorrect Numbers A and B) listed below:

Valid Number:	5703	1050	0151	9015
	↓	↓	↓	↓
Incorrect Number A:	$\underline{2}703$	$1\underline{4}50$	$01\underline{1}1$	$\underline{8}015$
Incorrect Number B:	$5\underline{0}03$	$1\underline{7}50$	$0\underline{8}51$	$\underline{2}015$

(a) Each member of the group chooses one of the valid numbers from the list above and performs the following tasks.

i. Compute the calculation to show that the number you chose is a valid identification number; that is, show that $a_4 = a_1a_2a_3 \pmod 7$.

ii. Consider the two single-digit errors associated with the valid number you chose. Apply the airline ticket check digit scheme to each of the incorrect numbers. Your calculations should indicate that the error involving incorrect number A will be caught while the error involving incorrect number B will not be caught.

iii. Take the valid number and subtract from it, in two separate calculations, the incorrect number A and the incorrect number B. For example, compute 5703 – 2703 and 5703 – 5003. Do you notice a difference between the two results? (HINT: Think about division by 7.)

(b) As a group, discuss your findings. Recall that if two numbers x and y are congruent modulo n, or $x \equiv y \pmod n$, that $x - y \equiv 0 \pmod n$. Use this idea, along with the results from part (a), to conjecture as to why the airline ticket will not catch the single-digit error when a is replaced by b, $| a - b |= 7$, and a is not the check digit.

Further Reading

Gallian, J. A., Check Digit Methods, *International Journal of Applied Engineering Education*, 5(4), 1989, 503–505.

Gallian, J. A., and Winters, S., Modular Arithmetic in the Marketplace, *American Mathematical Monthly*, 95, 1988, 548–551.

2.6 The Universal Product Code Check Digit Scheme

Preliminary Activity. Many of the products that are sold in stores have a 12-digit Universal Product Code (UPC) associated with them. Find four or five examples of 12-digit UPCs. Given the work on check digit schemes in the previous two sections, and the fact that the last of the 12 digits in a UPC is the check digit, attempt to determine the UPC check digit scheme.

There is a Universal Product Code (UPC) on every product in every store. UPCs are used in a variety of ways. First of all, a UPC identifies not only the product but also what type of product it is and its manufacturer. Using the UPC, each store specifies a price that appears on the cash register when the product's bar code is scanned as you check out. Stores also use UPCs for inventory purposes. By using the UPC, the store can track products that sell quickly, or ones that rarely sell, and make sure that all products are always in stock.

The UPC system was created in 1973 by the US grocery industry as the standard bar code for marking products. Three years later the European Article Numbering (EAN) code was developed. The latter is used by the rest of the world. There are five versions of the UPC and two of the EAN. The two versions most commonly used in the US are the 12-digit *Version A* UPC and the eight-digit *Version E* UPC systems. Version E is used in special cases, such as for soda cans, a variety of smaller items, and magazines. An example of each is given in Figure 2.2.

In this section, the 12-digit Version A system will be studied. In a 12-digit UPC a_1-$a_2a_3a_4a_5a_6$-$a_7a_8a_9a_{10}a_{11}$-a_{12}, the first digit a_1, referred to as the *number system character*, identifies the type of product. Table 2.1 lists the different values of a_1 and when each is used.

The second set of numbers $a_2a_3a_4a_5a_6$ identifies the manufacturer. The third set of numbers $a_7a_8a_9a_{10}a_{11}$ identifies the product. The last digit a_{12} is the check digit. While the check digit is not always printed, it is always encoded in the bar code. As companies grow and produce more and more, some have shortened their manufacturer number to four digits $a_2a_3a_4a_5$ and have lengthened their product identification number to six digits $a_6a_7a_8a_9a_{10}a_{11}$. Regardless of the style, the check digit scheme works the same [3], [4], [6], [29].

FIGURE 2.2
Left, Version A UPC; *right,* Version E UPC

TABLE 2.1
Character Values in Version A UPC

a_1	Specific Use
0	General groceries
2	Meat and produce
3	Drugs and health products
4	Non-food items
5	Coupons
6, 7	Other items
1, 8, 9	Reserved for future use

Definition 2.6.1. *Given the* 11*-digit number* a_1-$a_2a_3a_4a_5a_6$-$a_7a_8a_9a_{10}a_{11}$*, the check digit* a_{12} *is appended to create the UPC* a_1-$a_2a_3a_4a_5a_6$-$a_7a_8a_9a_{10}a_{11}$-a_{12} *such that* a_{12} *satisfies the equation*

$$3 \cdot a_1 + 1 \cdot a_2 + 3 \cdot a_3 + 1 \cdot a_4 + 3 \cdot a_5 + 1 \cdot a_6 + 3 \cdot a_7 + 1 \cdot a_8 + 3 \cdot a_9 + 1 \cdot a_{10} + 3 \cdot a_{11} + 1 \cdot a_{12} = 0 \pmod{10}.$$

The equation in Definition 2.6.1 is often written as follows:

$$(3,1,3,1,3,1,3,1,3,1,3,1) \cdot (a_1,a_2,a_3,a_4,a_5,a_6,a_7,a_8,a_9,a_{10},a_{11},a_{12}) = 0 \pmod{10}.$$

The expression

$$(3,1,3,1,3,1,3,1,3,1,3,1) \cdot (a_1,a_2,a_3,a_4,a_5,a_6,a_7,a_8,a_9,a_{10},a_{11},a_{12}),$$

which equals

$$3 \cdot a_1 + 1 \cdot a_2 + 3 \cdot a_3 + 1 \cdot a_4 + 3 \cdot a_5 + 1 \cdot a_6 + 3 \cdot a_7 + 1 \cdot a_8 + 3 \cdot a_9 + 1 \cdot a_{10} + 3 \cdot a_{11} + 1 \cdot a_{12},$$

is referred to as the *dot product* of the two components $(3,1,3,1,3,1,3,1,3,1,3,1)$ and $(a_1,a_2,a_3,a_4,a_5,a_6,a_7,a_8,a_9,a_{10},a_{11},a_{12})$.

The UPC 0-53600-10054-0, presented in Figure 2.2, is valid because the following calculation results in a true statement ($40 = 4 \cdot 10 + 0$):

$$\begin{aligned}
(3,1,3,1,3,1,3,1,3,1,3,1) \cdot (0,5,3,6,0,0,1,0,0,5,4,0) &= 0 \pmod{10} \\
3 \cdot 0 + 1 \cdot 5 + 3 \cdot 3 + 1 \cdot 6 + 3 \cdot 0 + 1 \cdot 0 + 3 \cdot 1 + 1 \cdot 0 + 3 \cdot 0 + 1 \cdot 5 + 3 \cdot 4 + 1 \cdot 0 &= 0 \pmod{10} \\
0 + 5 + 9 + 6 + 0 + 0 + 3 + 0 + 0 + 5 + 12 + 0 &= 0 \pmod{10} \\
40 &= 0 \pmod{10}.
\end{aligned}$$

The UPC 3-70501-09110-4 is invalid, because the calculation below results in a false statement ($39 = 3 \cdot 10 + 9$):

$$\begin{aligned}(3,1,3,1,3,1,3,1,3,1,3,1) \cdot (3,7,0,5,0,1,0,9,1,1,0,4) &= 0 \pmod{10}\\ 3\cdot 3+1\cdot 7+3\cdot 0+1\cdot 5+3\cdot 0+1\cdot 1+3\cdot 0+1\cdot 9+3\cdot 1+1\cdot 1 &\\ +3\cdot 0+1\cdot 4 &= 0 \pmod{10}\\ 9+7+0+5+0+1+0+9+3+1+0+4 &= 0 \pmod{10}\\ 39 &= 0 \pmod{10}.\end{aligned}$$

Once the 11-digit product number of an item is created, this system is used to generate the check digit. Consider the product identification number 5-02003-91562. The check digit C will be appended to this number to create the UPC 5-02003-91562-C, such that C satisfies the equation $(3,1,3,1,3,1,3,1,3,1,3,1) \cdot (5,0,2,0,0,3,9,1,5,6,2,C) = 0 \pmod{10}$. To find C, this equation is solved:

$$\begin{aligned}(3,1,3,1,3,1,3,1,3,1,3,1) \cdot (5,0,2,0,0,3,9,1,5,6,2,C) &= 0 \pmod{10}\\ 3\cdot 5+1\cdot 0+3\cdot 2+1\cdot 0+3\cdot 0+1\cdot 3+3\cdot 9+1\cdot 1+3\cdot 5+1\cdot 6 &\\ +3\cdot 2+1\cdot C &= 0 \pmod{10}\\ 15+0+6+0+0+3+27+1+15+6+6+C &= 0 \pmod{10}\\ 79+C &= 0 \pmod{10}.\end{aligned}$$

The only digit that solves this equation is $C = 1$ ($79 + 1 = 80 = 8 \cdot 10 + 0$). Thus the UPC is 5-02003-91562-1.

The UPC check digit scheme is better than the "mod 7" and "mod 9" schemes of the previous sections. Those schemes did not catch all single-digit errors, but the "mod 10" UPC scheme does. Look at the UPC 5-02003-91562-1 and the following single-digit errors.

Correct Number:	5-02003-91562-**1**	5-020**0**3-91562-1
	↓	↓
Incorrect Number:	5-02003-91562-**7**	5-020**9**3-91562-1

Both errors are caught. The single-digit error involving the check digit results in the following false statement:

$$\begin{aligned}(3,1,3,1,3,1,3,1,3,1,3,1) \cdot (5,0,2,0,0,3,9,1,5,6,2,7) &= 0 \pmod{10}\\ 3\cdot 5+1\cdot 0+3\cdot 2+1\cdot 0+3\cdot 0+1\cdot 3+3\cdot 9+1\cdot 1+3\cdot 5+1\cdot 6 &\\ +3\cdot 2+1\cdot 7 &= 0 \pmod{10}\\ 15+0+6+0+0+3+27+1+15+6+6+7 &= 0 \pmod{10}\\ 86 &= 0 \pmod{10}.\end{aligned}$$

Actually, $86 = 6 \pmod{10}$, not 0. The single-digit error involving the fifth digit also results in a false statement:

$$\begin{aligned}(3,1,3,1,3,1,3,1,3,1,3,1) \cdot (5,0,2,0,9,3,9,1,5,6,2,1) &= 0 \pmod{10}\\ 3\cdot 5+1\cdot 0+3\cdot 2+1\cdot 0+3\cdot 9+1\cdot 3+3\cdot 9+1\cdot 1+3\cdot 5+1\cdot 6 &\\ +3\cdot 2+1\cdot 1 &= 0 \pmod{10}\\ 15+0+6+0+27+3+27+1+15+6+6+1 &= 0 \pmod{10}\\ 107 &= 0 \pmod{10}.\end{aligned}$$

Actually, $107 = 7 \pmod{10}$, not 0.

It is easy to show why this scheme catches all single-digit errors. Let $a_1 \ldots a_i \ldots a_{12}$ be a UPC with $1 \leq i \leq n$. When a single-digit error occurs, $a_1 \ldots a_i \ldots a_{12}$ is transmitted as $a_1 \ldots b_i \ldots a_{12}$ where $a_i \neq b_i$ (a single-digit error where a_i is replaced by b_i).

Suppose this error is not caught. Then both

$$(3, 1, \ldots, 3, 1) \cdot (a_1, \ldots, a_i, \ldots, a_{12}) = 0 \pmod{10}$$

and

$$(3, 1, \ldots, 3, 1) \cdot (a_1, \ldots, b_i, \ldots, a_{12}) = 0 \pmod{10}.$$

Written in another way,

$$\begin{aligned}(3, 1, \ldots, 3, 1) \cdot (a_1, \ldots, a_i, \ldots, a_{12}) - (3, 1, \ldots, 3, 1) \cdot (a_1, \ldots, b_i, \ldots, a_{12})& \\ = 0 \pmod{10}.&\end{aligned}$$

There are two possibilities to consider: When computing the dot product $(3, 1, \ldots, 3, 1) \cdot (a_1, \ldots, a_i, \ldots, a_{12})$ and $(3, 1, \ldots, 3, 1) \cdot (a_1, \ldots, b_i, \ldots, a_{12})$, both a_i and b_i are multiplied either by the number 3 or by the number 1.

- **Case I.** In this case, both a_i and b_i are multiplied by the number 3. This results in

$$\begin{aligned}0 &= (3, 1, \ldots, 3, 1) \cdot (a_1, \ldots, a_i, \ldots, a_{12}) \\ &\quad - (3, 1, \ldots, 3, 1) \cdot (a_1, \ldots, b_i, \ldots, a_{12}) \pmod{10} \\ &= (3 \cdot a_1 + 1 \cdot a_2 + \cdots + 3 \cdot a_i + \cdots + 3 \cdot a_{11} + 1 \cdot a_{12}) \\ &\quad - (3 \cdot a_1 + 1 \cdot a_2 + \cdots + 3 \cdot b_i + \cdots + 3 \cdot a_{11} + 1 \cdot a_{12}) \pmod{10} \\ &= 3 \cdot a_1 + 1 \cdot a_2 + \cdots + 3 \cdot a_i + \cdots + 3 \cdot a_{11} + 1 \cdot a_{12} \\ &\quad - 3 \cdot a_1 - 1 \cdot a_2 - \cdots - 3 \cdot b_i - \cdots - 3 \cdot a_{11} - 1 \cdot a_{12} \pmod{10} \\ &= 3 \cdot a_i - 3 \cdot b_i \pmod{10} \\ &= 3 \cdot (a_i - b_i) \pmod{10}.\end{aligned}$$

  Thus $3 \cdot (a_i - b_i) = 0 \pmod{10}$. Since $a_i \neq b_i$, $a_i - b_i \neq 0$. Because 3 and 10 are relatively prime, $3 \cdot (a_i - b_i)$ will never be a multiple of 10. This is a contradiction. Thus the assumption that the error is not caught is false, and the error is caught.

- **Case II.** In this case, both a_i and b_i are multiplied by the number 1. This results in

$$\begin{aligned}0 &= (3, 1, \ldots, 3, 1) \cdot (a_1, \ldots, a_i, \cdots, a_{12}) \\ &\quad - (3, 1, \ldots, 3, 1) \cdot (a_1, \ldots, b_i, \ldots, a_{12}) \pmod{10} \\ &= (3 \cdot a_1 + 1 \cdot a_2 + \cdots + 1 \cdot a_i + \cdots + 3 \cdot a_{11} + 1 \cdot a_{12}) \\ &\quad - (3 \cdot a_1 + 1 \cdot a_2 + \cdots + 1 \cdot b_i + \cdots + 3 \cdot a_{11} + 1 \cdot a_{12}) \pmod{10} \\ &= 3 \cdot a_1 + 1 \cdot a_2 + \cdots + 1 \cdot a_i + \cdots + 3 \cdot a_{11} + 1 \cdot a_{12} \\ &\quad - 3 \cdot a_1 - 1 \cdot a_2 - \cdots - 1 \cdot b_i - \cdots - 3 \cdot a_{11} - 1 \cdot a_{12} \pmod{10} \\ &= 1 \cdot a_i - 1 \cdot b_i \pmod{10} \\ &= a_i - b_i \pmod{10}.\end{aligned}$$

  Thus $a_i - b_i = 0 \pmod{10}$. Since $a_i \neq b_i$, $a_i - b_i \neq 0$. Also, since both a_i and b_i are digits between 0 and 9, $a_i - b_i$ can never be a multiple of 10. This is a contradiction. Thus the assumption that the error is not caught is false, and the error is caught.

However, the UPC scheme does not catch all transposition-of-adjacent-digits errors. If a and b are two adjacent digits in a UPC, the transposition error $\ldots ab\ldots \to \ldots ba\ldots$ is not caught when $|a-b|=5$. Consider the following two such errors.

Correct Number:	5-02003-**91**562-1	**5-0**2003-91562-1
	↓	↓
Incorrect Number:	5-02003-**19**562-1	**0-5**2003-91562-1

The first error, with $a=9$ and $b=1$, would be caught, as $|a-b|=|9-1|=|8|=8\neq 5$. The second error, with $a=5$ and $b=0$, would not be caught, as $|a-b|=|5-0|=|5|=5$.

Exercises 2.6

1. The following two numbers are UPCs. Using the UPC check digit scheme, determine which is a valid number and which is an invalid number.

   0-70501-09110-4 1-23004-24410-7

2. The number 3-88207-29431 has been generated by a manufacturer to identify a product that is going to market. Using the UPC check digit scheme, assign a check digit to this number. Be sure to identify the check digit and the entire UPC.
3. List all the possible numbers that could be check digits for the UPC check digit scheme. Explain how you got your answer.

Group Activities 2.6

1. It was mentioned above that the UPC scheme does not catch all transposition-of-adjacent-digits errors. If a and b are two adjacent digits in a UPC, the transposition error $\ldots ab\ldots \to \ldots ba\ldots$ is not caught when $|a-b|=5$. The goal of this activity is to determine why this happens. Consider the following four valid UPCs and their associated transposition-of-adjacent-digits errors. Each member of the group chooses one valid UPC to investigate.

Correct Number	Incorrect Number A	Incorrect Number B
5-02003-91562-1	5-02003-91652-1	0-52003-91562-1
1-03271-00082-0	1-30271-00082-0	1-03721-00082-0
2-90014-30161-7	2-90014-30167-1	2-90014-30611-7
4-72011-10391-7	4-70211-10391-7	4-27011-10391-7

   (a) Making sure to write out all the calculations, show that the Correct Number is actually a valid UPC.

   (b) Making sure to write out all the calculations, show that the transposition error in Incorrect Number A will be caught by the UPC scheme, while the transposition error in Incorrect Number B will not be caught.

(c) Show that for Incorrect Number A, $\mid a - b \mid \neq 5$, where a and b are the two digits in the number that are transposed. Show that for Incorrect Number B, $\mid a - b \mid = 5$, where a and b are the two digits in the number that are transposed.

(d) In parts (a) and (b), the following dot product was computed for each number, where $a_1a_2a_3a_4a_5a_6a_7a_8a_9a_{10}a_{11}a_{12}$ was the correct or incorrect UPC:

$$(3,1,3,1,3,1,3,1,3,1,3,1)\cdot(a_1,a_2,a_3,a_4,a_5,a_6,a_7,a_8,a_9,a_{10},a_{11},a_{12}) = 0 \pmod{10}.$$

i. Compare the dot product calculation for the Correct Number with the one for Incorrect Number A. Now compare it with the one for Incorrect Number B. In both cases, you should see only a minor difference between the pairs of calculations. What is it in each case?

ii. For both cases (Correct and Incorrect A; Correct and Incorrect B), subtract the two calculations from each other. What do you get? How does the value in each case compare to the number 10? Recall that the UPC is a "mod 10" scheme.

(e) Given your work in the previous steps, conjecture as to why the transposition of adjacent digits a and b (that is, $\ldots ab \ldots \to \ldots ba \ldots$) is not caught when $\mid a - b \mid = 5$.

2. For a 12-digit UPC a_1-$a_2a_3a_4a_5a_6$-$a_7a_8a_9a_{10}a_{11}$-a_{12}, the dot product of $(3,1,3,1,3,1,3,1,3,1,3,1)$ and $(a_1,a_2,a_3,a_4,a_5,a_6,a_7,a_8,a_9,a_{10},a_{11},a_{12})$ is taken to determine whether the UPC is valid or not. The goal of this activity is to determine why the positive integers 3 and 1 are chosen. Recall that the UPC scheme catches all single-digit errors.

(a) Suppose that the UPC scheme assigned the check digit a_{12} such that

$$(2,1,2,1,2,1,2,1,2,1,2,1)\cdot(a_1,a_2,a_3,a_4,a_5,a_6,a_7,a_8,a_9,a_{10},a_{11},a_{12}) = 0 \pmod{10}.$$

This switching to $(2,1,2,1,2,1,2,1,2,1,2,1)$ will affect the UPC scheme, and it will no longer be able to catch all single-digit errors. Consider these four valid UPCs and their associated single-digit errors. Each member of the group chooses one valid UPC to investigate.

Correct Number	IncorrectNumber A	Incorrect Number B
0-45723-38152-7	5-45723-38152-7	0-45773-38152-7
1-03271-00082-3	1-08271-00082-3	1-03221-00082-3
2-90014-30161-5	2-90014-30166-5	2-90014-30661-5
4-72011-10391-9	4-77011-10391-9	4-72011-10891-9

i. Making sure to write out all the calculations, show that the Correct Number is actually a valid UPC.

ii. Making sure to write out all the calculations, show that the single-digit errors in Incorrect Numbers A and B are not caught by the new UPC scheme.

iii. Compare your calculation on the valid UPC with the calculation involving Incorrect Number A and with the one involving Incorrect Number B. Each pair of calculations should differ at only one spot. What is it in each case?

iv. For both cases (Correct and Incorrect A; Correct and Incorrect B), subtract the two calculations from each other. What do you get? How does each result compare to the number 10? Recall that the UPC is a "mod 10"scheme.

v. Note that the single-digit error in each incorrect number differed from the corresponding digit in the valid UPC by 5. This difference was then multiplied by 2, causing the error not to be caught. Would multiplying by a different number (say 3) have caused the error to be caught? What is the relationship between 2 and 10? What is the relationship between 3 and 10? (HINT: Think in terms of division.)

(b) Now suppose that the UPC scheme assigned the check digit a_{12} such that

$$(5,1,5,1,5,1,5,1,5,1,5,1) \cdot (a_1, a_2, a_3, a_4, a_5, a_6, a_7, a_8, a_9, a_{10}, a_{11}, a_{12}) = 0 \pmod{10}.$$

This switching to $(5,1,5,1,5,1,5,1,5,1,5,1)$ will affect the UPC scheme and it will no longer be able to catch all single-digit errors. Consider these four valid UPCs and their associated single-digit errors. Each member of the group chooses one valid UPC to investigate.

Correct Number	Incorrect Number A	Incorrect Number B
6-81320-03711-0	6-83320-03711-0	6-81360-03711-0
1-03271-00082-4	3-03271-00082-4	1-03251-00082-4
2-90014-30161-1	2-90014-30163-1	2-90014-50161-1
4-72011-10391-3	4-74011-10391-3	4-72011-90391-3

i. Making sure to write out all the calculations, show that the Correct Number is actually a valid UPC.

ii. Making sure to write out all the calculations, show that the single-digit errors in Incorrect Numbers A and B are not caught by the new UPC scheme.

iii. Compare your calculation performed on the valid UPC with the calculation involving Incorrect Number A and with the one involving Incorrect Number B. Each pair of calculations should differ at only one spot. What is it in each case?

iv. For both cases (Correct and Incorrect A; Correct and Incorrect B), subtract the two calculations from each other. What do you get? In each case how does the value compare to the number 10? Recall that the UPC is a "mod 10" scheme.

v. Note that the single-digit error in each incorrect number differed from the corresponding digit in the valid UPC by 2. This difference was then multiplied by 5, causing the error not to be caught. Would multiplying by a different number (say 3) have caused the error to be caught? What is the relationship between 5 and 10? Between 3 and 10? (HINT: Think in terms of division.)

(c) Based on the two activities above, conjecture why the numbers 3 and 1 are used in the UPC scheme. Are there other numbers that would work just as well?

Further Reading

Gallian, J. A., The Mathematics of Identification Numbers, *College Mathematics Journal*, 22(3), 1991, 194–202.

Gallian, J. A., Error Detection Methods, *ACM Computing Surveys*, 28(3), 1996, 504–517.

UPC Symbol Specification Manual, Uniform Code Council, Dayton, Ohio, 1986.

Wood, E. F., Self-Checking Codes: An Application of Modular Arithmetic, *Mathematics Teacher*, 80, 1987, 312–316.

2.7 The International Standard Book Number Check Digit Scheme

Preliminary Activity. All books have a ten-digit ISBN associated with them, which is usually printed on the back cover. Find four or five examples of ISBNs. Given the work on check digit schemes in the previous three sections and the fact that the last of the ten digits in an ISBN is the check digit, attempt to determine the ISBN check digit scheme.

Starting in the years 1968–1972, all books published were assigned a ten-digit number: an International Standard Book Number (ISBN). This system was developed as book publishers and wholesalers began to computerize their inventories. Identifying a book by its author, title, edition, and so on, was replaced by an internationally recognized number. This made book ordering easier and more efficient, overcame language barriers (ordering books from foreign countries), and cleared up many other problems.

The presentation of an ISBN varies slightly from book to book [12], [25]. While each ISBN is always ten digits long, the arrangement of the digits can differ slightly. The first number in an ISBN is the *group* or *country* number, and it identifies the language area and the nation or the geographic grouping of the publisher. For example, 0 indicates that the book was published in the English-speaking world (United Kingdom, United States, Australia, etc.), 2 indicates the book was published in the French-speaking world (Belgium, France, etc.), 3 indicates it was published in the German-speaking world (Austria, Germany, etc.), the number 87 indicates Denmark, and the number 90 indicates Holland. The second number in an ISBN identifies the publisher. This number usually is from two to five digits in length but can be longer.

Example 2.7.1. *Each of the following ISBNs begins with a* 0, *indicating that all were published in an English-speaking country.*

- 0-19-853287-3 is the ISBN for the book *Codes and Cryptography* by Dominic Welsh [28]. The publisher code is 19 and denotes Oxford University Press.
- 0-471-51001-7 is the ISBN for the book *Modern Algebra: An Introduction* by John Durbin [2]. The publisher code is 471 and denotes John Wiley & Sons, Inc.

- 0-8176-3805-9 is the ISBN for the book *Math into LaTeX: An Introduction to LaTeX and AMS-LaTeX* by George Grätzer [9]. The publisher code is 8176 and denotes Birkhäuser.
- 0-86720-498-2 is the ISBN for the book *Introduction to Linear Algebra* by Géza Schay [21]. The publisher code is 86720 and denotes Jones and Bartlett Publishers.

The third number in the ISBN is the code that the publisher has chosen for the book. Note in the examples above that a longer publisher code leaves fewer digits available for book titles. Thus a publisher who has a large number of book titles will want a shorter publisher code. The last number is always a single digit and is the check digit.

For example, the following two ISBNs are used to identify Neal Koblitz's book *A Course in Number Theory and Cryptography* [14]:

0-387-96576-9 3-540-96576-9

In the first ISBN, the leading 0 indicates the version of the book published in English. The second part of this ISBN, 387, is the code for the Springer-Verlag publishing company in New York City. The third part, 96576, identifies the book and is the same for both versions. The last digit, 9, is the check digit. In the second ISBN, the leading 3 indicates the version of the book published in German. The second part of this ISBN, 540, is the code for the Springer-Verlag publishing company in Berlin. The third part, 96576, identifies the book and is the same for both versions. The last digit, 9, is the check digit.

Definition 2.7.2. The ISBN Check Digit Scheme [4], [25]. *For the ten-digit ISBN, denoted* $a_1a_2a_3a_4a_5a_6a_7a_8a_9a_{10}$*, the check digit* a_{10} *that is appended to the nine-digit number* $a_1a_2a_3a_4a_5a_6a_7a_8a_9$ *is determined by the equation*

$$10 \cdot a_1 + 9 \cdot a_2 + 8 \cdot a_3 + 7 \cdot a_4 + 6 \cdot a_5 + 5 \cdot a_6 + 4 \cdot a_7 + 3 \cdot a_8 + 2 \cdot a_9 + 1 \cdot a_{10} = 0 \pmod{11}.$$

If the check digit a_{10} *is* 10*, the letter* X *is used instead.*

The equation presented above in Definition 2.7.2 is often written as follows:

$$(10, 9, 8, 7, 6, 5, 4, 3, 2, 1) \cdot (a_1, a_2, a_3, a_4, a_5, a_6, a_7, a_8, a_9, a_{10}) = 0 \pmod{11}.$$

The expression $(10, 9, 8, 7, 6, 5, 4, 3, 2, 1) \cdot (a_1, a_2, a_3, a_4, a_5, a_6, a_7, a_8, a_9, a_{10})$, which equals

$$10 \cdot a_1 + 9 \cdot a_2 + 8 \cdot a_3 + 7 \cdot a_4 + 6 \cdot a_5 + 5 \cdot a_6 + 4 \cdot a_7 + 3 \cdot a_8 + 2 \cdot a_9 + 1 \cdot a_{10},$$

is the dot product of $(10, 9, 8, 7, 6, 5, 4, 3, 2, 1)$ and $(a_1, a_2, a_3, a_4, a_5, a_6, a_7, a_8, a_9, a_{10})$.

The remainder when a number is divided by 11 could be any digit from 0 to 9 or the number 10. Since the ISBN scheme uses modulo 11 arithmetic and requires the check digit a_{10} to be a single character, it assigns a_{10} the value of X when 10 is the check digit. The ISBN for the book *Linear Algebra and Its Applications* by David Lay [15] is 0-201-52032-X. The X indicates that the check digit is the number 10. This is a valid number, as the following calculations indicate:

$$
\begin{aligned}
(10,9,8,7,6,5,4,3,2,1)\cdot(0,2,0,1,5,2,0,3,2,X) &= 0 \pmod{11}\\
(10,9,8,7,6,5,4,3,2,1)\cdot(0,2,0,1,5,2,0,3,2,10) &= 0 \pmod{11}\\
10\cdot 0+9\cdot 2+8\cdot 0+7\cdot 1+6\cdot 5+5\cdot 2+4\cdot 0+3\cdot 3+2\cdot 2 &\\
+1\cdot 10 &= 0 \pmod{11}\\
0+18+0+7+30+10+0+9+4+10 &= 0 \pmod{11}\\
88 &= 0 \pmod{11}.
\end{aligned}
$$

This is a true statement ($88 = 8\cdot 11+0$). An example of an invalid number would be 3-357-02001-4. This can be seen from the following calculations:

$$
\begin{aligned}
(10,9,8,7,6,5,4,3,2,1)\cdot(3,3,5,7,0,2,0,0,1,4) &= 0 \pmod{11}\\
10\cdot 3+9\cdot 3+8\cdot 5+7\cdot 7+6\cdot 0+5\cdot 2+4\cdot 0+3\cdot 0+2\cdot 1+1\cdot 4 &= 0 \pmod{11}\\
30+27+40+49+0+10+0+0+2+4 &= 0 \pmod{11}\\
162 &= 0 \pmod{11}.
\end{aligned}
$$

This is a false statement, as $162 = 8 \pmod{11}$ ($162 = 14\cdot 11 + 8$).

To create an ISBN for a book with an identification number of 0-3021-9041, we need to append the check digit C, creating the ISBN 0-3021-9041-C. We must choose a valid C so that $(10,9,8,7,6,5,4,3,2,1)\cdot(0,3,0,2,1,9,0,4,1,C) = 0 \pmod{11}$. First we calculate

$$
\begin{aligned}
(10,9,8,7,6,5,4,3,2,1)\cdot(0,3,0,2,1,9,0,4,1,C) &= 0 \pmod{11}\\
10\cdot 0+9\cdot 3+8\cdot 0+7\cdot 2+6\cdot 1+5\cdot 9+4\cdot 0+3\cdot 4+2\cdot 1 &\\
+1\cdot C &= 0 \pmod{11}\\
0+27+0+14+6+45+0+12+2+C &= 0 \pmod{11}\\
106+C &= 0 \pmod{11}.
\end{aligned}
$$

Since $106 = 7\cdot 11+0$, the smallest nonnegative number C that can be added to 106 to make $106+C$ a multiple of 11 will be $C=0$. Thus the check digit $C=0$ and the ISBN is 0-3021-9041-0.

As shown in Table 1.2, over 90% of all errors are single-digit and transposition-of-adjacent-digits errors. Given this fact, a check digit scheme ought to catch, at the very least, these two types of errors. None of the schemes investigated in previous sections are able to do so. But the ISBN scheme *does* catch all single-digit and transposition-of-adjacent-digits errors.

It is easy to show why this scheme catches all single-digit errors. Let $a_1\dots a_i\dots a_{10}$ be an ISBN with $1\le i\le n$. When a single-digit error occurs, $a_1\dots a_i\dots a_{10}$ is transmitted as $a_1\dots b_i\dots a_{10}$ with $a_i\ne b_i$ (a single-digit error where a_i is replaced by b_i).

Suppose this error is not caught. Then both

$$(10,\dots,1)\cdot(a_1,\dots,a_i,\dots,a_{10}) = 0 \pmod{11}$$

and

$$(10,\dots,1)\cdot(a_1,\dots,b_i,\dots,a_{10}) = 0 \pmod{11}.$$

This can also be written as

$$(10,\dots,1)\cdot(a_1,\dots,a_i,\dots,a_{10}) - (10,\dots,1)\cdot(a_1,\dots,b_i,\dots,a_{10}) = 0 \pmod{11}.$$

When we compute the dot product $(10, \ldots, 1) \cdot (a_1, \ldots, a_i, \ldots, a_{10})$ and the dot product $(10, \ldots, 1) \cdot (a_1, \ldots, b_i, \ldots, a_{10})$, both a_i and b_i will be multiplied by the same integer k, where $1 \leq k \leq 10$. Since 11 is a prime number, k will be relatively prime to 11, no matter what its value. Now calculate

$$\begin{aligned} 0 &= (10, \ldots, k, \ldots, 1) \cdot (a_1, \ldots, a_i, \ldots, a_{10}) \\ &\qquad - (10, \ldots, k, \ldots, 1) \cdot (a_1, \ldots, b_i, \ldots, a_{10}) \pmod{11} \\ &= (10 \cdot a_1 + \cdots + k \cdot a_i + \cdots + 1 \cdot a_{10}) \\ &\qquad - (10 \cdot a_1 + \cdots + k \cdot b_i + \cdots + 1 \cdot a_{10}) \pmod{11} \\ &= 10 \cdot a_1 + \cdots + k \cdot a_i + \cdots + 1 \cdot a_{10} \\ &\qquad - 10 \cdot a_1 - \cdots - k \cdot b_i - \cdots - 1 \cdot a_{10} \pmod{11} \\ &= k \cdot a_i - k \cdot b_i \pmod{11} \\ &= k \cdot (a_i - b_i) \pmod{11}. \end{aligned}$$

Thus $k \cdot (a_i - b_i) = 0 \pmod{11}$. Now, since $a_i \neq b_i$, and both are digits between 0 and 9, $a_i - b_i \neq 0$ and $-9 \leq a_i - b_i \leq 9$. Since k and 11 are relatively prime, $k \cdot (a_i - b_i)$ will never be a multiple of 11. This is a contradiction. Thus the assumption that the error is not caught is false, and the error is caught.

Although the ISBN scheme does have the advantage of catching all single-digit and transposition-of-adjacent-digits errors, it has two drawbacks. First of all, it uses the letter X to represent the case when the check digit is 10. It is preferable to have the check digit be one of the digits $0, 1, 2, 3, 4, 5, 6, 7, 8$, or 9. This avoids the introduction of new characters in identification numbers that are made up entirely of digits. Secondly, while this is a fairly successful scheme (in terms of catching transmission errors), it only works with identification numbers of length 10.

This gives us a new goal to shoot for: To find a check digit scheme that catches all single-digit and transposition-of-adjacent-digits errors, does not introduce any new characters into the identification number system, and can be used with identification numbers of any length.

Exercises 2.7

1. Using the ISBN check digit scheme, determine which of the following is a valid ISBN and which is invalid.

   3-824-27519-X $\qquad\qquad$ 0-387-94461-3

2. The number 5-283-11980 has been generated by a publisher to identify a book that it plans to market. Using the ISBN check digit scheme, assign a check digit to this number. Be sure to identify the check digit and the entire ISBN number.

3. Recall the publishers mentioned in Example 2.7.1, and suppose for the moment that each one publishes books in the English-speaking world only. How many possible book code numbers are available to them?

   (a) Oxford University Press
   (b) John Wiley & Sons, Inc.
   (c) Birkhäuser
   (d) Jones and Bartlett Publishers

Paper Assignment 2.7

1. **Comparison.** You are employed by an organization that wants to adopt a check digit scheme. At this point, you are familiar with only two groups of schemes: Group I consists of the US postal and airline ticket check digit schemes, and Group II consists of the UPC and ISBN check digit schemes. Your job is to recommend, in a report called a feasibility study, either a scheme from Group I or a scheme from Group II. To do this, compare Group I with Group II, paying attention to the following factors:

   (a) ease of use (the length of, time involved with, and complexity of the calculations);

   (b) strength (in terms of the errors it can detect).

   In making your evaluation, consider also the type of organization that will employ the check digit scheme.

Further Reading

Gallian, J. A., The Mathematics of Identification Numbers, *College Mathematics Journal*, 22(3), 1991, 194–202.

Gallian, J. A., Error Detection Methods, *ACM Computing Surveys*, 28(3), 1996, 504–517.

The ISBN System Users' Manual, International ISBN Agency, Berlin, 1986.

Tuchinsky, P. M., International Standard Book Numbers, *The UMAP Journal*, 5, 1989, 41–54.

Wood, E. F., Self-Checking Codes: An Application of Modular Arithmetic, *Mathematics Teacher*, 80, 1987, 312–316.

2.8 Cryptography and the RSA Public-Key System

Several number theoretic concepts (prime numbers, modulo arithmetic, relatively prime, etc.) were introduced in the first three sections of this chapter. Many of them are central to the creation and implementation of check digit schemes. They also can be applied in the area of cryptography to create some fairly sophisticated codes that are extremely difficult to break. The RSA Public-Key Cryptosystem, discussed in this section, is one such code.

Preliminary Activity. Each group develops and writes out its own alphanumeric code for the purpose of sending secret messages. The code does not need to be fancy or complex. As a group, write a message using the code you created, then trade messages with another group. Attempt to break the code and figure out the other group's secret message.

Consider the following situation: Alice wants to send Bob a message. The only problem is that Alice and Bob do not want anyone else to be able to read the message, and they know that a third person, Cindy, is viewing all of their correspondence. The only choice that Alice and Bob have, to keep Cindy from reading all of their messages, is to develop a code that Cindy cannot understand and then to send messages using that code.

The development of this code brings a host of new problems. First, Alice and Bob need to decide on a message coding system. Second, this coding system needs to be difficult to crack, or Cindy will break the code and read all of the messages. And finally, Alice and Bob need to exchange the *keys*, which are the methods for creating and reading the coded messages. This exchange must be done in secret, without Cindy's knowledge. Since Cindy is listening in on all of Alice and Bob's communications, they cannot send the coding method to each other. The safest method would then be for Alice and Bob to meet. However, this exchange is a problem, because Alice and Bob are hundreds of miles apart.

Historical Notes. Traditionally, only the military and the diplomatic service needed to send coded messages. Today, however, the sending of coded messages extends far beyond governmental agencies. With the advent of electronic communication, electronic banking, and accessing and sending data over the Internet, corporations, banks, shoppers, and many others want to insure that the information and messages sent are read only by the intended recipients. Mathematicians have taken the lead in developing these secure communication techniques.

How to create, send, and read coded messages is an age-old problem that dates back to the ancient societies of Egypt, Mesopotamia, India, and China. In 400 B.C., the Spartans used a form of secret writing called the Scytale. The person sending a message first wrapped a piece of parchment around a cylinder and then wrote the message on the parchment, being sure to write one letter on each revolution of the parchment strip. The parchment was then unwrapped and sent. To anyone who did not know Scytale, the parchment showed a list of seemingly meaningless letters. The receiver of the message would read it by rewrapping the parchment around a cylinder of the same diameter.

Julius Caesar is said to have developed a secret writing system so that secure messages could be sent back and forth between himself and his generals in the field. Today, armies all over the world routinely send coded messages for the same reason. During World War II, two major Allied victories occurred when the Allied forces broke Japan's Purple code [16], [18] and Germany's Enigma code [8], [13], [27]. Once the Allies broke the Purple code, they knew about the movements of Japan's military in the Pacific. Enigma was actually a coding machine developed by the German inventor Arthur Scheribus in the early 1920s. It had a typewriter keyboard and rotary letter wheels which coded the message. Portable field units were used by all branches of the German military in an attempt to secure all their communications. When the Allies broke the Enigma code, they were able to read all of Germany's military messages.

Definition 2.8.1. ***Cryptography*** *is the study of methods to send messages in disguised form so that only the intended recipient can decode and read the message.*

The message that is to be sent is called the *plaintext* and the message in coded form is called the *ciphertext*. There are two main tasks involved in sending a coded message:

1. Taking the plaintext and rewriting it in ciphertext.
2. Taking the ciphertext message and obtaining the plaintext message from it.

Definition 2.8.2. *The process of taking plaintext and writing it in ciphertext is called* ***enciphering*** *or* ***encryption****. The process of taking a ciphertext message and obtaining the plaintext message from it is called* ***deciphering****. The method used to encipher and decipher messages is called a* ***cipher****.*

There are many different ciphers. The value of the cipher depends on how easy it is to crack (i.e., determine the code being used). With the introduction of computers, ciphers have had to become more complex so that they are harder to break.

Plaintext Shift or Caesar Ciphers. The simplest ciphers are *plaintext shift ciphers* or *Caesar Ciphers* (so named because they were used by Julius Caesar [23]). This type of cipher is created by uniformly shifting the letters in the alphabet over a certain number of spaces.

For example, suppose that Alice wants to send Bob the message "*The Berlin wall will fall tonight.*" First, Alice and Bob must agree on the plaintext shift they will use. They agree to use the one presented in Figure 2.3, which shifts the alphabet over nine spaces.

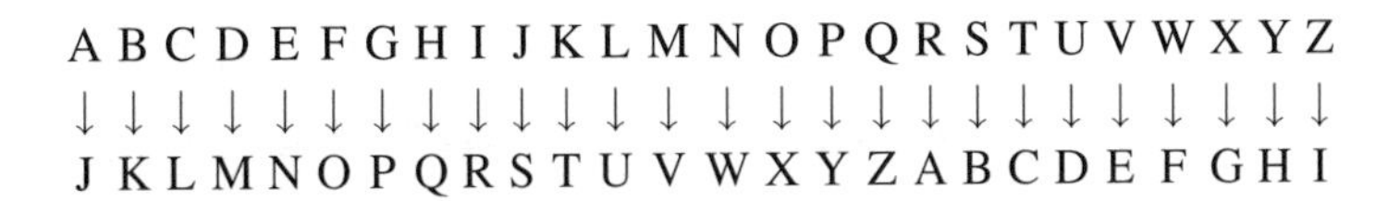

FIGURE 2.3
Nine-Space Plaintext Shift

To encipher the message, Alice takes each letter in the message, finds it in the top row of Figure 2.3, reads down the arrow, and replaces it with the corresponding letter on the bottom row. The result is presented in Figure 2.4.

THE	BERLIN	WALL	WILL	FALL	TONIGHT.
↓	↓	↓	↓	↓	↓
CQN	KNAURW	FJUU	FRUU	OJUU	CXWRPQC.

FIGURE 2.4
Sample Nine-Space Plaintext Shift Message

Alice then sends the ciphertext "*CQN KNAURW FJUU FRUU OJUU CXWRPQC.*" When Bob receives this message, he takes each letter in the message, locates it in the bottom row of Figure 2.3, reads up (in the opposite direction of the arrow), and replaces it with the corresponding letter on the top row. By doing this he gets back the plaintext message.

Substitution or Monoalphabetic Ciphers. A variation of the plaintext shift is the *substitution* or *monoalphabetic* cipher. This method involves permuting the letters of the alphabet. In other words, the letters are more than simply shifted: they are jumbled up into an arbitrary order. For example, suppose that Alice wants to send Bob the message "*The Berlin wall will fall tonight.*" First, Alice and Bob must agree on the monoalphabetic cipher they will use. They agree to use the one presented in Figure 2.5.

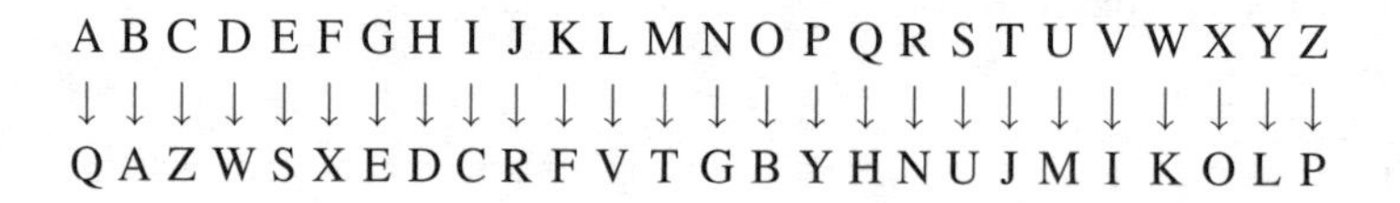

FIGURE 2.5
An Example of a Substitution Cipher

To encipher the message, Alice takes each letter in the message, finds it in the top row of Figure 2.5, reads down the arrow, and replaces it with the corresponding letter on the bottom row. By doing this for each letter in the message, the ciphertext is obtained, as illustrated in Figure 2.6.

THE	BERLIN	WALL	WILL	FALL	TONIGHT.
↓	↓	↓	↓	↓	↓
JDS	ASNVCG	KQVV	KCVV	XQVV	JBGCEDJ.

FIGURE 2.6
Sample Ciphertext Message Using a Substitution Cipher

Alice then sends the ciphertext

JDS ASNVCG KQVV KCVV XQVV JBGCEDJ

When Bob receives this message, he takes each letter in the message, finds it in the bottom row of Figure 2.5, reads up (in the opposite direction of the arrow), and replaces it with the corresponding letter on the top row. By doing this he gets the plaintext message back.

While these two ciphers work, they are not reliable as they can be broken very easily. This is where mathematics gets involved. Mathematics can be used to create sophisticated ciphers that are virtually impossible to break. To begin this discussion, we will first see how a cipher, very similar to a plaintext shift or Caesar cipher, can be obtained mathematically by using some of the number theoretic concepts introduced earlier in this chapter.

Plaintext Shift or Caesar Ciphers Revisited. Recall the plaintext shift cipher presented earlier (see Figure 2.7). This cipher motivates the use of mathematics in cryptography.

A	B	C	D	E	F	G	H	I	J	K	L	M	N	O	P	Q	R	S	T	U	V	W	X	Y	Z
↓	↓	↓	↓	↓	↓	↓	↓	↓	↓	↓	↓	↓	↓	↓	↓	↓	↓	↓	↓	↓	↓	↓	↓	↓	↓
J	K	L	M	N	O	P	Q	R	S	T	U	V	W	X	Y	Z	A	B	C	D	E	F	G	H	I

FIGURE 2.7
A Sample Plaintext Shift Cipher

To start things off, assign a number to each letter of the alphabet, as shown in Table 2.2. Assign A the integer 00, B the integer 01, C the integer 02, and so on until Z, which is assigned the integer 25. Each integer is called the letter's *numerical equivalent*. The space between words also has a numerical equivalent, namely 26. The numbers $0, 1, 2, 3, 4, 5, 6, 7, 8$, and 9 are written as $00, 01, 02, 03, 04, 05, 06, 07, 08$, and 09 so that each letter corresponds to a two-digit number. The reason for assigning the letter A to the integer 0 and not 1 will be explained later.

TABLE 2.2
Numerical Equivalents

A	B	C	D	E	F	G	H	I	J	K	L	M	N	O	P	Q	R	S	T	U	V	W	X	Y	Z	*space*
↓	↓	↓	↓	↓	↓	↓	↓	↓	↓	↓	↓	↓	↓	↓	↓	↓	↓	↓	↓	↓	↓	↓	↓	↓	↓	↓
00	01	02	03	04	05	06	07	08	09	10	11	12	13	14	15	16	17	18	19	20	21	22	23	24	25	26

Now recall the plaintext shift presented in Figure 2.7. To generate a ciphertext message, the letter A changes to J, the letter B changes to K, the letter C changes to L, the letter Q changes to Z, and so on. Rewriting this process using each letter's numerical equivalent would result in 00 (A) being changed 09 (J), 01 (B) being changed to 10 (K), 02 being changed to 11 (L), 16 (Q) being changed to 25 (Z), and so on. This correspondence can be achieved by adding 9 to each of the numerical equivalents ($00 + 9 = 09$, $01 + 9 = 10$, $02 + 9 = 11$, and $16 + 9 = 25$).

But simply adding 9 is not enough. The plaintext shift present in Figure 2.7 changes V, whose numerical equivalent is 21, to E, with numerical equivalent 04. Adding 9 to 21 yields the number 30, but no letter corresponds to 30. So, how can we achieve the "wrap-around" effect of the plaintext shift, if adding 9 is not enough? Adding 9 and computing the result modulo 27 will work. This results in the following mathematically based cipher, which is approximately the same[4] as the plaintext shift in Figure 2.7.

1. Take each letter of the message and assign it a two-digit number P using Table 2.2.
2. Take each number P and encipher it to the number C using the rule

$$C = (P + 9) \pmod{27}.$$

[4] It will not be exactly the same, because this new cipher assigns the space between words the number 26. The space between words was not taken into account in the original plaintext shift cipher.

3. To decipher the message, take each two-digit number C and get back the numerical equivalent P by using the rule

$$P = (C + 18) \pmod{27}.$$

4. Generate the plaintext message from the numerical equivalents P using Table 2.2.

Now it can seen why the numerical equivalents start with the number 0 or 00. The cipher mentioned above assigns the digit C to be the remainder when P is divided by 27. Since the remainders when any number is divided by 27 are $0, 1, 2, \ldots, 26$, the numerical equivalents make sense.

Suppose Alice wants to send Bob the message "*Sell all shares.*" She first assigns numerical equivalents to all the letters and spaces in the message, as shown in the second row of Table 2.3. Alice then takes each numerical equivalent P and computes $(P + 9) \pmod{27}$, the third row of Table 2.3.

TABLE 2.3
Enciphering the Plaintext Message "Sell All Shares"

S	e	l	l		a	l	l		s	h	a	r	e	s
↓	↓	↓	↓	↓	↓	↓	↓	↓	↓	↓	↓	↓	↓	↓
18	04	11	11	26	00	11	11	26	18	07	00	17	04	18
↓	↓	↓	↓	↓	↓	↓	↓	↓	↓	↓	↓	↓	↓	↓
00	13	20	20	08	09	20	20	08	00	16	09	26	13	00

For example, the letter S is first assigned the numerical equivalent $P = 18$. To get the number C, Alice performs the following calculation:

$$\begin{aligned} C &= (P + 9) \pmod{27} \\ &= (18 + 9) \pmod{27} \\ &= 27 \pmod{27} \\ &= 00. \end{aligned}$$

Continuing in this manner for each numerical equivalent P, Alice sends the message

001320200809202008001609261300.

Bob then receives the message, breaks it up into two-digit pairs, and deciphers the message using the procedure in steps 3 and 4 above. This process is presented in Table 2.4.

To decipher the message, Bob uses the rule $P = (C + 18) \pmod{27}$. For example, the first two-digit pair is 00. Bob calculates $C = 00$ as follows:

$$\begin{aligned} P &= (C + 18) \pmod{27} \\ &= (00 + 18) \pmod{27} \\ &= 18 \pmod{27} \\ &= 18. \end{aligned}$$

TABLE 2.4
Deciphering the Ciphertext

00	13	20	20	08	09	20	20	08	00	16	09	26	13	00
↓	↓	↓	↓	↓	↓	↓	↓	↓	↓	↓	↓	↓	↓	↓
18	04	11	11	26	00	11	11	26	18	07	00	17	04	18
↓	↓	↓	↓	↓	↓	↓	↓	↓	↓	↓	↓	↓	↓	↓
S	e	l	l		a	l	l		s	h	a	r	e	s

In the same way, for the third two-digit pair, Bob calculates $C = 20$:

$$\begin{aligned} P &= (C + 18) \pmod{27} \\ &= (20 + 18) \pmod{27} \\ &= 38 \pmod{27} \\ &= 11. \end{aligned}$$

Since 18 corresponds to the letter S and 11 corresponds to the letter L, Bob now knows the first and third letters of the message.

The RSA Public-Key Cryptosystem. All of the ciphers presented so far can easily be broken. A person who has intercepted a message encrypted with one of these ciphers could just try every possible letter or number (00 to 26) permutation to decipher it. This is a very simple task, since a computer can test all of the possibilities in a matter of minutes. Such easy cracking defeats the purpose of creating the cipher in the first place. One that is not easily broken is needed.

Furthermore, the "meeting" problem has not yet been solved. With every cipher studied so far, Alice and Bob have to meet or communicate somehow so that they can agree on what cipher to use. If Cindy is listening to all of their conversations, she will find out how to decipher their messages.

Both of these problems are addressed in the RSA public-key cryptosystem developed by R. L. Rivest, A. Shamir, and L. Adleman [19], [20]. Although it is a fairly simple cipher, it is very hard to break. Moreover, Alice and Bob do not need to meet and agree on a cipher. The code is based on modulo arithmetic, and Alice, the sender of the message, only needs to know how to encipher a message, while Bob, the receiver, is the only person who needs to know how to decipher it. The cipher works as follows.

1. The receiver finds two distinct large prime numbers, p and q, and lets the integer $n = p \cdot q$. Normally, these primes are each over 100 digits long. The larger the primes are, the harder it is to break the code.
2. The receiver finds a large integer r that is relatively prime to the integer $m = (p-1) \cdot (q-1)$.
3. The receiver finds the unique integer s such that $1 \le s \le m$ and $s \cdot r \equiv 1 \pmod{m}$.

   There are several methods that enable the receiver to compute s very quickly. One method is presented here. Since r and m are relatively prime, there is a positive integer k such that $r^k \equiv 1 \pmod{m}$. Given that $r^k = r^{k-1} \cdot r$, $r^k \equiv 1 \pmod{m}$

results in $r^{k-1} \cdot r \equiv 1 \pmod{m}$. Consequently, once this value of k is found, $s = r^{k-1} \pmod{m}$ will satisfy $s \cdot r \equiv 1 \pmod{m}$. To find k, compute $r \pmod{m}$, $r^2 \pmod{m}$, $r^3 \pmod{m}$, $r^4 \pmod{m}$, and so on until the positive integer k is found such that $r^k \pmod{m} = 1$.

4. The receiver makes the pairs of numbers r and n public. Anyone (even Cindy) can have access to these numbers because they are used only for encryption and do not enable anyone to break the code. They are known as the *public key*.
5. The sender of the message takes the numerical equivalent (using Table 2.2) of the message to be sent and apportions it into a sequence of numbers, such that each number M in the sequence has the same number of digits and $1 \leq M \leq n$.
6. The sender then enciphers each M into ciphertext C by using the rule

$$C = M^r \pmod{n}.$$

7. The receiver is sent the ciphertext and uses the *private key* s and n to decipher C. To get the number M from C, the receiver uses the rule

$$M = C^s \pmod{n}.$$

Here are two examples of how the RSA cipher works. Recall that Alice is the sender of the message and Bob is the receiver. In Example 2.8.3, very small prime numbers will be used so that the calculations are not difficult. In Example 2.8.4, the primes are still relatively small, but the calculations are much more involved.

Example 2.8.3.

1. Bob picks two prime numbers, $p = 5$ and $q = 7$, then computes

$$n = p \cdot q = 5 \cdot 7 = 35.$$

2. Bob computes

$$m = (p - 1) \cdot (q - 1) = (5 - 1) \cdot (7 - 1) = 4 \cdot 6 = 24.$$

Now he must find an integer r that is relatively prime to $m = 24$. The prime factorization of $24 = 2^3 \cdot 3$, so $r = 5$ is relatively prime to $m = 24$.

3. Bob must find the unique integer s such that $1 \leq s \leq m$ and

$$s \cdot r \equiv 1 \pmod{m}$$

or

$$s \cdot 5 \equiv 1 \pmod{24}.$$

In other words, he must find s such that $s \cdot 5$ has a remainder of 1 when divided by 24.

To find s, Bob first must find k such that $r^k \equiv 1 \pmod{m}$. In this case, Bob has $5^k \equiv 1 \pmod{24}$. So he starts by computing $r^2 \pmod{24}$, $r^3 \pmod{24}$, $r^4 \pmod{24}$, and so on, with $r = 5$, until he finds a number k such that $r^k \pmod{24} = 1$. At this point he stops and $s = r^{k-1} \pmod{24}$.

Bob begins as follows:

$$r^2 \pmod{24} = 5^2 \pmod{24} = 25 \pmod{24} = 1.$$

Since $5^2 \pmod{24} = 1$, he stops and lets $k = 2$.

Consequently,

$$s = r^{k-1} \pmod{24} = 5^{2-1} \pmod{24} = 5^1 \pmod{24} = 5 \pmod{24} = 5,$$

or $s = 5$. In other words, $s = 5$ is the unique number such that $s \cdot r = 5 \cdot 5$ has a remainder of 1 when divided by 24:

$$\begin{aligned} s \cdot r &= 5 \cdot 5 \\ &= 25 \\ &= 1 \cdot 24 + 1. \end{aligned}$$

4. Bob makes the pair of numbers $r = 5$ and $n = 35$ public. They are the *public key*. In particular, he gives them to Alice so that she can send him a coded message. Cindy can also have these numbers. They are used only to encode messages, not to decode them.
5. Alice wants to send Bob the message "Cindy is a spy." She first takes the plaintext message and then, using Table 2.2, converts the message into its numerical equivalence:

$$\begin{array}{ccccccc} \text{Cindy} & & \text{is} & & \text{a} & & \text{spy} \\ \downarrow & \downarrow & \downarrow & \downarrow & \downarrow & \downarrow & \downarrow \\ 0208130324 & 26 & 0818 & 26 & 00 & 26 & 181524 \end{array}$$

The numerical equivalence of the message is 0208130324260818260026181524. Since this is a very large number, much larger than $n = 35$, the message needs to be broken down into a sequence of smaller numbers, such that each of the smaller numbers has the same number of digits and each is not larger than 35. Alice breaks down the large number into a sequence of numbers, each with two digits:

02 08 13 03 24 26 08 18 26 00 26 18 15 24

6. Alice now takes each number M from this sequence and enciphers it into ciphertext C by using the rule

$$\begin{aligned} C &= M^r \pmod{n} \\ &= M^5 \pmod{35}. \end{aligned}$$

If the resulting number C has fewer than two digits, enough zeros are added before the number to make a total of two digits. Since this sequence contains 14 numbers, let

$$\begin{array}{ccccc} M_1 = 02, & M_2 = 08, & M_3 = 13, & M_4 = 03, & M_5 = 24, \\ M_6 = 26, & M_7 = 08, & M_8 = 18, & M_9 = 26, & M_{10} = 00, \\ M_{11} = 26, & M_{12} = 18, & M_{13} = 15, & \text{and} & M_{14} = 24. \end{array}$$

The specific calculations follow:

$$\begin{aligned}
C_1 &= M_1^5 \pmod{35} = (02)^5 \pmod{35} = 32 \pmod{35}\\
C_2 &= M_2^5 \pmod{35} = (08)^5 \pmod{35} = 32768 \pmod{35}\\
C_3 &= M_3^5 \pmod{35} = (13)^5 \pmod{35} = 371293 \pmod{35}\\
C_4 &= M_4^5 \pmod{35} = (03)^5 \pmod{35} = 243 \pmod{35}\\
C_5 &= M_5^5 \pmod{35} = (24)^5 \pmod{35} = 7962624 \pmod{35}\\
C_6 &= M_6^5 \pmod{35} = (26)^5 \pmod{35} = 11881376 \pmod{35}\\
C_7 &= M_7^5 \pmod{35} = (08)^5 \pmod{35} = 32768 \pmod{35}\\
C_8 &= M_8^5 \pmod{35} = (18)^5 \pmod{35} = 1889568 \pmod{35}\\
C_9 &= M_9^5 \pmod{35} = (26)^5 \pmod{35} = 11881376 \pmod{35}\\
C_{10} &= M_{10}^5 \pmod{35} = (00)^5 \pmod{35} = 0 \pmod{35}\\
C_{11} &= M_{11}^5 \pmod{35} = (26)^5 \pmod{35} = 11881376 \pmod{35}\\
C_{12} &= M_{12}^5 \pmod{35} = (18)^5 \pmod{35} = 1889568 \pmod{35}\\
C_{13} &= M_{13}^5 \pmod{35} = (15)^5 \pmod{35} = 759375 \pmod{35}\\
C_{14} &= M_{14}^5 \pmod{35} = (24)^5 \pmod{35} = 7962624 \pmod{35}.
\end{aligned}$$

As a result,

$$\begin{aligned}
C_1 &= 32 \pmod{35} = 32 && 32 = 0 \cdot 35 + 32;\\
C_2 &= 32768 \pmod{35} = 08 && 32768 = 936 \cdot 35 + 8;\\
C_3 &= 371293 \pmod{35} = 13 && 371293 = 10608 \cdot 35 + 13;\\
C_4 &= 243 \pmod{35} = 33 && 243 = 6 \cdot 35 + 33;\\
C_5 &= 7962624 \pmod{35} = 19 && 7962624 = 227503 \cdot 35 + 19;\\
C_6 &= 11881376 \pmod{35} = 31 && 11881376 = 339467 \cdot 35 + 31;\\
C_7 &= 32768 \pmod{35} = 08 && 32768 = 936 \cdot 35 + 8;\\
C_8 &= 1889568 \pmod{35} = 23 && 1889568 = 53987 \cdot 35 + 23;\\
C_9 &= 11881376 \pmod{35} = 31 && 11881376 = 339467 \cdot 35 + 31;\\
C_{10} &= 0 \pmod{35} = 0 && 0 = 0 \cdot 35 + 0;\\
C_{11} &= 11881376 \pmod{35} = 31 && 11881376 = 339467 \cdot 35 + 31;\\
C_{12} &= 1889568 \pmod{35} = 23 && 1889568 = 53987 \cdot 35 + 23;\\
C_{13} &= 759375 \pmod{35} = 15 && 759375 = 21696 \cdot 35 + 15;\\
C_{14} &= 7962624 \pmod{35} = 19 && 7962624 = 227503 \cdot 35 + 19.
\end{aligned}$$

Thus the ciphertext that Alice sends to Bob is

C_1	C_2	C_3	C_4	C_5	C_6	C_7	C_8	C_9	C_{10}	C_{11}	C_{12}	C_{13}	C_{14}
32	08	13	33	19	31	08	23	31	00	31	23	15	19

7. Bob receives this ciphertext message and uses the *private key* $s = 5$ and $n = 35$ to decipher it. To get each number M from each number C in the ciphertext, Bob uses

the rule

$$M = C^s \pmod{n}$$
$$M = C^5 \pmod{35}.$$

The calculations are similar to those above; only the results will be shown here:

$$\begin{aligned}
M_1 &= C_1^5 \pmod{35} = (32)^5 \pmod{35} = 02\\
M_2 &= C_2^5 \pmod{35} = (08)^5 \pmod{35} = 08\\
M_3 &= C_3^5 \pmod{35} = (13)^5 \pmod{35} = 13\\
M_4 &= C_4^5 \pmod{35} = (33)^5 \pmod{35} = 03\\
M_5 &= C_5^5 \pmod{35} = (19)^5 \pmod{35} = 24\\
M_6 &= C_6^5 \pmod{35} = (31)^5 \pmod{35} = 26\\
M_7 &= C_7^5 \pmod{35} = (08)^5 \pmod{35} = 08\\
M_8 &= C_8^5 \pmod{35} = (23)^5 \pmod{35} = 18\\
M_9 &= C_9^5 \pmod{35} = (31)^5 \pmod{35} = 26\\
M_{10} &= C_{10}^5 \pmod{35} = (00)^5 \pmod{35} = 00\\
M_{11} &= C_{11}^5 \pmod{35} = (31)^5 \pmod{35} = 26\\
M_{12} &= C_{12}^5 \pmod{35} = (23)^5 \pmod{35} = 18\\
M_{13} &= C_{13}^5 \pmod{35} = (15)^5 \pmod{35} = 15\\
M_{14} &= C_{14}^5 \pmod{35} = (19)^5 \pmod{35} = 24.
\end{aligned}$$

Bob now has the numerical equivalent of the message:

02 08 13 03 24 26 08 18 26 00 26 18 15 24

or 0208130324260818260026181524. He then uses Table 2.2 to get the plaintext message "Cindy is a spy":

02	08	13	03	24	26	08	18	26	00	26	18	15	24
↓	↓	↓	↓	↓	↓	↓	↓	↓	↓	↓	↓	↓	↓
C	i	n	d	y		i	s		a		s	p	y

Example 2.8.4.

1. Bob picks two prime numbers $p = 37$ and $q = 73$. He then computes

$$n = p \cdot q = 37 \cdot 73 = 2701.$$

2. Bob computes

$$m = (p-1)\cdot(q-1) = (37-1)\cdot(73-1) = 36 \cdot 72 = 2592$$

and now must find an integer r that is relatively prime to $m = 2592$. The prime factorization of $2592 = 2^5 \cdot 3^4$, so $r = 7$ is relatively prime to $m = 2592$.

3. Now Bob must find the unique integer s such that $1 \leq s \leq m$ and

$$s \cdot r \equiv 1 \pmod{m} \qquad \text{or} \qquad s \cdot 7 \equiv 1 \pmod{2592}.$$

In other words, he must find s such that $s \cdot 7$ has a remainder of 1 when divided by 2592.

To find s, Bob first must find k such that $r^k \equiv 1 \pmod{m}$, or in this case such that $7^k \equiv 1 \pmod{2592}$. He starts by computing $r^2 \pmod{2592}$, $r^3 \pmod{2592}$, $r^4 \pmod{2592}$, and so on with $r = 7$, until he finds a number k such that $r^k \pmod{2592} = 1$. Then $s = r^{k-1} \pmod{2592}$. The modulo arithmetic techniques introduced earlier help simplify this task:

- $r^2 \pmod{2592} = 7^2 \pmod{2592} = 49 \pmod{2592} = 49$;
- $r^3 \pmod{2592} = 7^3 \pmod{2592} = 343 \pmod{2592} = 343$;
- $r^4 \pmod{2592} = 7^4 \pmod{2592} = 2401 \pmod{2592} = 2401$;
- $r^5 \pmod{2592} = 7^5 \pmod{2592} = 16807 \pmod{2592} = 1255$;
- $r^6 \pmod{2592} = 7^6 \pmod{2592} = 117649 \pmod{2592} = 1009$;
- continue in this manner until
- $r^{108} \pmod{2592} = 7^{108} \pmod{2592} = 1$.

Hence $k = 108$. Thus

$$s = r^{k-1} \pmod{2592} = 7^{108-1} \pmod{2592} = 7^{107} \pmod{2592} = 1111,$$

or $s = 1111$. In other words, $s = 1111$ is the unique number such that $s \cdot r = 1111 \cdot 7$ has a remainder of 1 when divided by 2592:

$$\begin{aligned} s \cdot r &= 1111 \cdot 7 \\ &= 7777 \\ &= 3 \cdot 2592 + 1. \end{aligned}$$

4. Bob makes the pair of numbers $r = 7$ and $n = 2701$ public. They are the *public key*. In particular, he gives them to Alice so that she can send him a coded message.
5. Alice wants to send Bob the message "Cindy is a spy." She takes the plaintext message and then, using Table 2.2, converts the message into its numerical equivalence:

Cindy		is		a		spy
↓	↓	↓	↓	↓	↓	↓
0208130324	26	0818	26	00	26	181524

The numerical equivalent of the message is 0208130324260818260026181524. Since this is a very large number, much larger than $n = 2701$, the message needs to be broken down into a sequence of smaller numbers, each of which has the same number of digits and is not larger than 2701. Alice breaks down the large number into a sequence of numbers, each with four digits:

0208 1303 2426 0818 2600 2618 1524

6. Alice now takes each number M from this sequence and enciphers it into ciphertext

C by using the rule

$$\begin{aligned} C &= M^r \pmod{n} \\ &= M^7 \pmod{2701}. \end{aligned}$$

If the resulting number C has fewer than four digits, enough zeros are added before the number to make it a total of four digits. Since this sequence contains seven numbers, let $M_1 = 0208, M_2 = 1303, M_3 = 2426, M_4 = 0818, M_5 = 2600, M_6 = 2618$, and $M_7 = 1524$. The modulo arithmetic techniques introduced earlier help simplify the calculations. Only the results of each calculation are given here:

$$\begin{aligned}
C_1 &= M_1^7 \pmod{2701} = (0208)^7 \pmod{2701} = 1420 \\
C_2 &= M_2^7 \pmod{2701} = (1303)^7 \pmod{2701} = 325 \\
C_3 &= M_3^7 \pmod{2701} = (2426)^7 \pmod{2701} = 2023 \\
C_4 &= M_4^7 \pmod{2701} = (0818)^7 \pmod{2701} = 2583 \\
C_5 &= M_5^7 \pmod{2701} = (2600)^7 \pmod{2701} = 1157 \\
C_6 &= M_6^7 \pmod{2701} = (2618)^7 \pmod{2701} = 2241 \\
C_7 &= M_7^7 \pmod{2701} = (1524)^7 \pmod{2701} = 2254.
\end{aligned}$$

Thus the ciphertext that Alice sends to Bob is

C_1	C_2	C_3	C_4	C_5	C_6	C_7
1420	0325	2023	2583	1157	2241	2254

7. Bob now receives this ciphertext message and uses the *private key* $s = 1111$ to decipher it. To get each number M from each number C in the ciphertext, Bob uses the rule

$$\begin{aligned} M &= C^s \pmod{n} \\ M &= C^{1111} \pmod{2701}. \end{aligned}$$

This will be a lengthy calculation, as each number must be raised to the 1111th power. Computers are normally used to do these lengthy calculations. However, the modulo arithmetic techniques introduced earlier can be used to simplify the calculations. Only the results of the calculations are given here:

$$\begin{aligned}
M_1 &= C_1^{1111} \pmod{2701} = (1420)^{1111} \pmod{2701} = 0208 \\
M_2 &= C_2^{1111} \pmod{2701} = (0325)^{1111} \pmod{2701} = 1303 \\
M_3 &= C_3^{1111} \pmod{2701} = (2023)^{1111} \pmod{2701} = 2426 \\
M_4 &= C_4^{1111} \pmod{2701} = (2583)^{1111} \pmod{2701} = 0818 \\
M_5 &= C_5^{1111} \pmod{2701} = (1157)^{1111} \pmod{2701} = 2600 \\
M_6 &= C_6^{1111} \pmod{2701} = (2241)^{1111} \pmod{2701} = 2618 \\
M_7 &= C_7^{1111} \pmod{2701} = (2254)^{1111} \pmod{2701} = 1524.
\end{aligned}$$

Bob now has the numerical equivalent of the message:

0208 1303 2426 0818 2600 2618 1524

or 0208130324260818260026181524. He then uses Table 2.2 to get the plaintext message "Cindy is a spy":

02	08	13	03	24	26	08	18	26	00	26	18	15	24
↓	↓	↓	↓	↓	↓	↓	↓	↓	↓	↓	↓	↓	↓
C	i	n	d	y		i	s		a		s	p	y

Exercises 2.8

1. The coded message "YMJ HTIJ MFX GJJS GWTPJS" has been intercepted. A plaintext shift (Caesar cipher) was used to generate the message. Determine the specific plaintext shift used and what the plaintext message is.
2. Consider the nine-space plaintext shift cipher depicted in Figure 2.3.
   (a) What ciphertext message would Alice send to Bob if her plaintext message is "Iraq still has chemical weapons"?
   (b) What plaintext message did Alice send to Bob if he received the ciphertext "LRWMH RB J BYH"?
3. How many plaintext shift ciphers are there? Explain how you arrived at your answer.
4. Consider the substitution cipher depicted in Figure 2.5.
   (a) What ciphertext message would Alice send to Bob if her plaintext message is "Iraq still has chemical weapons"?
   (b) What plaintext message did Alice send to Bob if he received the ciphertext "LBMN XNCSGW JBT CU Q WBMAVS QESGJ"?
5. How many monoalphabetic ciphers are there? Explain how you arrived at your answer.
6. Consider the following cipher.
   - For each letter and space in the plaintext message, write its numerical equivalent P using Table 2.2.
   - Obtain each ciphertext letter C by computing $C = P + 15 \pmod{27}$.
   - Decode each ciphertext letter C into its original numerical equivalent P by computing $P = C + 12 \pmod{27}$.
   - Using Table 2.2, take the numerical equivalents P and obtain the original plaintext message.

   (a) What ciphertext message would Alice send to Bob if her plaintext message is "Iraq still has chemical weapons"? Remember that Alice must find the numerical equivalence of this message using Table 2.2 and then generate the ciphertext.
   (b) What plaintext message did Alice send to Bob if he received this ciphertext?

   180214010207140705080607141723011812

7. Recall the RSA public-key cipher presented in Example 2.8.3.

(a) What ciphertext message would Alice send to Bob if her plaintext message is "Call Abe"? Remember that Alice must find the numerical equivalence of this message using Table 2.2, break up the message into a sequence of numbers, each two digits long, and then generate the ciphertext.

(b) What plaintext message did Alice send to Bob if he received this ciphertext?

$$3200161631000109$$

8. Recall the RSA public-key cipher presented in Example 2.8.4. What ciphertext message would Alice send to Bob if her plaintext message is "Call"? Remember that Alice must find the numerical equivalence of this message using Table 2.2, break up the message into a sequence of numbers, each four digits long, and then generate the ciphertext.

9. Suppose that you want to create an RSA public-key cipher so that you and your friends can communicate in privacy over the Internet. The two primes you pick to be the basis for the cipher are $p = 7$ and $q = 11$. Find the numbers r and n that all people sending messages in ciphertext would need, and the numbers s and n that all receivers would need to decipher messages into plaintext.

Paper Assignments 2.8

1. **Serializing.** There are numerous other ciphers besides those presented here. The goal of this assignment is for you to find and present a cipher other than one presented in this section. The focus of this assignment is on describing a process in logical order (i.e., serializing). First present the cipher, and then list and describe the steps that must be followed to encode, send, and decipher messages. Your essay should include the following parts.

   (a) Explain clearly, with a few examples, how to encode a simple message. Provide the reader with the illustrations necessary to follow the encoding process (e.g., charts, tables).

   (b) Indicate how the message is to be sent. For example, will the sender send a huge number, a sequence of smaller numbers, or a sequence of letters?

   (c) Using your examples from part (a), explain clearly how to decipher a simple message. Again, provide the reader with the illustrations necessary to follow the deciphering process.

2. **Serializing.** The goal of this assignment is for you to develop your own cipher which Internet users can employ to send secure coded messages to one another through email. The focus here is on describing a process in logical order (i.e., serializing). First establish what cipher will be used, and then present the steps that must be followed to encode, send, and decipher messages.

   Explain how the sender and receiver will establish, agree on, and exchange all necessary information on the cipher. Caution: If they do this through the Internet, someone else may intercept their messages. In addition, if the sender lives in New York and the receiver in Kuala Lumpur, it will be hard for them to meet face to face. Your paper should have the following parts.

(a) Explain clearly, giving a few examples, how to encode a simple message. Provide the reader with the illustrations necessary to follow the encoding process (e.g., charts, tables).

(b) Indicate how messages will be sent over the Internet. For example, will the sender send a huge number, a sequence of smaller numbers, or a sequence of letters?

(c) Using your examples from part (a), explain clearly how to decipher a simple message. Again, provide the reader with the illustrations necessary to follow the deciphering process.

3. **Collaborative Writing.** Complete the writing assignment presented in (2) above in collaboration with a classmate.

Further Reading

Beutelspacher, A., *Cryptology*, The Mathematical Association of America, Washington, D.C., 1994.

Garliènski, J., *The Enigma War,* Scribner, New York, 1979.

Jones, C., *Navajo Code Talkers: Native American Heroes,* Tudor Publishers, Greensboro, 1997.

Kawano, K., et al., *Warriors: Navajo Code Talkers,* Northland Publishing, Flagstaff, 1990.

Khan, D., *Seizing the Enigma: The Race to Break the German U-Boat Codes, 1939–1943,* Houghton-Mifflin, Boston, 1991.

Kahn, D., *The Codebreakers: The Comprehensive History of Secret Communications from Ancient Times to the Internet,* Scribner, New York, 1996.

Lee, B., *Marching Orders: The Untold Story of World War II,* Crown Publishers, New York, 1995.

Lewin, R., *The American Magic: Codes, Ciphers, and the Defeat of Japan,* Farrar Straus Giroux, New York, 1982.

Luciano, D., and Prichett, G., Cryptology: From Caesar Ciphers to Public-Key Cryptosystems, *College Mathematics Journal*, 18(1), 1987, 2–17.

Newton, D. E., *Encyclopedia of Cryptology*, ABC-CLIO, Santa Barbara, 1997.

Prados, J., *Combined Fleet Decoded: The Secret History of American Intelligence and the Japanese Navy in World War II,* Random House, New York, 1995.

Welchman, G., *The Hut Six Story: Breaking the Enigma Codes,* McGraw Hill, New York, 1982.

Wrixon, F. B., *Codes and Ciphers,* Prentice Hall, New York, 1992.

Zimmermann, R. R., Cryptography for the Internet, *Scientific American*, 279(4), 1998, 110–115.

3

Functions, Permutations, and Their Applications

For each check digit scheme discussed in Chapter 2, the emphasis was on how the scheme caught errors when a number was transmitted. However, there is another important issue to consider: how identification numbers are created. Recall that an identification number is a string of digits, letters, symbols, or a combination of them that is used to identify a specific product, account, document, individual, and so on. The process used to create a UPC for a product and an ISBN for a book has already been discussed.

Before investigating identification numbers, it is necessary to discuss the mathematical concept of a set. This is important because the process of creating a single identification number to identify an item motivates a discussion of functions and permutations, which requires a firm knowledge of sets. Furthermore, permutations will be used to create check digit schemes more sophisticated than the ones presented in Chapter 2. One of these schemes, developed by IBM, is presented in this chapter.

3.1 Sets

Preliminary Activity. Search in print or electronic media to find a well-defined collection of objects (i.e., a set) from any field outside of mathematics. Identify the common characteristics that define the elements as members of the set. Are specific subcategories of this set of special interest? Write down your observations and bring them to class.

In Chapter 1, counting numbers, whole numbers, integers, rational numbers, and real numbers were defined. Each one was a specific collection of numbers satisfying a certain property. However, they are not the only collections of interest in mathematics; there are many more. A collection of objects that satisfies a certain condition or conditions is called a *set*. Sets can be manipulated and are central to a variety of topics that are presented in this chapter and the rest of the book.

Definition 3.1.1. *A **set** is a well-defined collection of objects called **elements** or **members**.*

To denote a set, the elements or members of the set are listed between the left and right braces $\{$ and $\}$. To refer to a specific set, capital letters are used (A, B, C, etc.). Elements from a specific set are denoted by lowercase letters (a, b, c, etc.). If a is an element of the set A, this is denoted by $a \in A$. For example, the set A of prime numbers between 1 and 10 is $A = \{2, 3, 5, 7\}$. Put another way, $2 \in A$, $3 \in A$, $5 \in A$, and $7 \in A$.

It does not matter in what order the elements of a set are listed. For example,

$$A = \{2, 3, 5, 7\} = \{3, 5, 7, 2\} = \{5, 7, 2, 3\} = \{7, 2, 3, 5\}.$$

If an element is in a set, it needs to be listed only once. Listing it twice or more is redundant. The set $\{2, 2, 2, 3, 3, 5, 7, 7\}$ and the set $\{2, 3, 3, 3, 5, 5, 7\}$ both equal the set $A = \{2, 3, 5, 7\}$.

Sets can be comprised of things other than numbers. The set

$$B = \{red, orange, yellow, green, blue, indigo, violet\}$$

is the collection of colors in the rainbow. Here, $red \in B$, $orange \in B$, and so on.

Saying that a set is a *well-defined collection of objects* means that however the set is denoted or presented, it is clear which objects are members of the set and which objects are not. In the examples above, it is clear which integers are and which integers are not members of set A, and which colors are and which colors are not members of set B. For example, 13 is not in the set A. While it is a prime number, it is not between 1 and 10. To denote this, $13 \notin A$ is written. Similarly, $pink \notin B$.

A set can have a finite number or an infinite number of elements. When possible, all the elements of the set should be listed, as in sets A and B above. However, when the set has a large or infinite number of elements, the ellipsis notation $\ldots$ is used to denote the pattern of elements in the set.

Example 3.1.2.

- Let C be the set of integers between 1 and 500. We can denote it by $C = \{1, 2, 3, 4, \ldots, 500\}$, indicating that all the integers between 4 and 500 are also in the set.
- Let D be the infinite set of all positive even integers. This is denoted $D = \{2, 4, 6, 8, 10, \ldots\}$, indicating that the pattern established by the first numbers in the set is continued and that all positive even integers are in D.

A set need not have any elements in it. The *empty set*, denoted $\{\}$ or $\emptyset$, is the set that has no elements in it.

Definition 3.1.3. *The **order** or **cardinality** of a finite set A, denoted $\mid A \mid$, is the number of elements in the set.*

Example 3.1.4. *Consider the finite sets*

$$\begin{aligned} A &= \{2, 3, 5, 7\}, \\ B &= \{red, orange, yellow, green, blue, indigo, violet\}, \\ C &= \{1, 2, 3, 4, \ldots, 500\}. \end{aligned}$$

- $\mid A \mid= 4$, $\mid B \mid= 7$, and $\mid C \mid= 500$.
- The order of the empty set is 0 ($\mid \emptyset \mid= 0$), as there are no elements in it.

Definition 3.1.5. *The set A is a **subset** of a set B, denoted $A \subseteq B$, if every element of A is also an element of B.*

Example 3.1.6. *Consider the sets*

$$\begin{aligned} A &= \{0,2,4,6,8\}, \\ B &= \{0,1,2,3,4,5,6,7,8,9\}, \\ C &= \{0,1,2,3,4,5\}. \end{aligned}$$

- $A \subseteq B$ and $C \subseteq B$, because each element that is in A and C is also in B.
- However, A is not a subset of C, denoted $A \not\subseteq C$. While the elements 0, 2, and 4 of set A are in set C, the elements 6 and 8 of A are not.

A set usually has many subsets. Given a set A, the empty set $\emptyset$ is always a subset. If $\mid A \mid\geq 1$, then the set A itself is another subset. The set $B = \{1,4\}$ has four subsets: $\emptyset$, $\{1\}$, $\{4\}$, and $\{1,4\}$.

Sets can also be combined to create new ones.

Definition 3.1.7. *Let A and B be sets. The **union** of A and B, denoted $A \cup B$, is the set of elements that are contained in either A or B.*

Definition 3.1.8. *Let A and B be sets. The **intersection** of A and B, denoted $A \cap B$, is the set of elements that are contained in both A and B.*

Example 3.1.9. *Consider the sets*

$$\begin{aligned} A &= \{0,1,2,3,4\}, \\ B &= \{3,4,5,6,7\}, \\ C &= \{1,2,4,6,8,10\}, \\ D &= \{10,11,12,13\}. \end{aligned}$$

- $A \cup B = \{0,1,2,3,4,5,6,7\}$.
- $B \cup C = \{1,2,3,4,5,6,7,8,10\}$.
- $A \cup B \cup D = \{0,1,2,3,4,5,6,7,10,11,12,13\}$.
- $A \cap B = \{3,4\}$.
- $A \cap C = \{1,2,4\}$.
- $B \cap D = \emptyset$. (There are no elements that are in both B and D).

Definition 3.1.10. *A **partition** of a finite set A is a collection of subsets $A_1, A_2, \ldots, A_r$ of A such that*

1. *$A_1 \cup A_2 \cup \cdots \cup A_r = A$, and*
2. *$A_i \cap A_j = \emptyset$ for all $1 \leq i \neq j \leq r$. In other words, given any two distinct subsets, no element is contained in both of them.*

Example 3.1.11. *Consider the set* $A = \{0, 1, 2, 3, 4, 5, 6, 7, 8, 9\}$.

- $A_1 = \{0, 1, 4, 8\}$, $A_2 = \{3\}$, $A_3 = \{5, 7, 9\}$, and $A_4 = \{2, 6\}$ form a partition of A, since $A_1 \cup A_2 \cup A_3 \cup A_4 = A$ and no two subsets have any elements in common.
- $A_1 = \{1, 4\}, A_2 = \{3, 5\}$, and $A_3 = \{0, 7, 9\}$ do not form a partition of A, as $A_1 \cup A_2 \cup A_3 = \{0, 1, 3, 4, 5, 7, 9\}$, which does not equal A.
- $A_1 = \{1, 4, 5, 8\}$, $A_2 = \{3\}$, $A_3 = \{0, 9\}$, and $A_4 = \{2, 5, 7\}$ do not form a partition of A, as $A_1 \cap A_4 = \{5\} \neq \emptyset$.

Exercises 3.1

1. Consider the set $A = \{0, 2, 4, 6, 8, 10, 12, 14, 16, 18, 20\}$.
   (a) Is $12 \in A$?
   (b) Is $5 \in A$?
   (c) What is the order or cardinality of the set A?
2. Give an example of a finite set and an infinite set. If it is impossible to list all the elements in one of these sets, be sure to clearly indicate (using any of methods used in this section) what the members of the set are.
3. Describe in a sentence or two the elements of the following sets.
   (a) $A = \{1, 3, 5, 7, 9, 11, \dots\}$
   (b) $B = \{2, 3, 5, 7, 11, 13, 17, 19, \dots, 127\}$
   (c) $C = \{\dots, -7, -5, -3, -1, 1, 3, 5, 7, \dots\}$
4. This exercise can be completed as homework or as Group Activity 3.1.1.
   (a) List all the subsets of each of these sets, keeping in mind that every set always has the empty set and itself as subsets: $\emptyset$; $\{0\}$; $\{0, 1\}$; $\{0, 1, 2\}$; $\{0, 1, 2, 3\}$.
   (b) Compute the following powers of 2: 2^0, 2^1, 2^2, 2^3, and 2^4.
   (c) Given the work you have done in parts (a) and (b), do you see a relationship between the number of elements in a set and the number of its subsets? Could you perform a simple calculation to find the number of subsets a finite set will have? Consider a set A with $\mid A \mid= n$. How many subsets will set A have? Test your hypothesis by using it to predict the number of subsets of the set $\{0, 1, 2, 3, 4\}$.
5. Given $A = \{1, 5, 10\}$, $B = \{2, 5, 9\}$. $C = \{2, 3, 4, 5, 6, 7, 8\}$, and $D = \{6, 7, 8, 9\}$, find the following sets.
   (a) $A \cap B$ (b) $B \cup D$ (c) $C \cap D$ (d) $B \cup C$
6. Identify the following statements as true or false. You need to indicate how you got your answer. Saying true or false is not enough.
   (a) $5 \in \{2, 4, 5, 8\}$ (b) $5 \subseteq \{2, 4, 5, 8\}$

(c) $\{5\} \subseteq \{2,4,5,8\}$

(d) $\{1,2\} \in \{\{1,3\},\{2,3\},\{1,2\}\}$

(e) $\{1,2\} \subseteq \{\{1,3\},\{2,3\},\{1,2\}\}$

(f) $\emptyset \in \{\emptyset\}$

(g) $\emptyset \subseteq \{\emptyset\}$

7. This exercise can be completed as homework or as Group Activity 3.1.2.

   (a) Find all the partitions of each of these sets: $\{0\}$, $\{0,1\}$, $\{0,1,2\}$.

   (b) For an integer $n \geq 1$, $n!$ (read *n factorial*) is the product $n\cdot(n-1)\cdot(n-2)\cdots 2\cdot 1$. For example, $5! = 5\cdot 4\cdot 3\cdot 2\cdot 1 = 120$. Compute $1!$, $2!$, and $3!$.

   (c) Given your work in parts (a) and (b), do you see a relationship between the number of elements in a set and the number of partitions of that set? Could you perform a simple calculation to find the number of subsets a finite set will have? Consider a set A with $|A| = n$. How many partitions will set A have? Test your hypothesis by using it to predict the number of partitions of the set $\{0,1,2,3\}$.

8. Find all the partitions of the set $A = \{2,5,8,9\}$.

9. Consider two finite sets A and B such that $A \cup B = A$.

   (a) Give three examples of sets that satisfy this property.

   (b) In each of the three examples generated in part (a), how does set A relate to set B? Describe this relationship or condition using terms introduced in this section.

   (c) In general, what can one say about set B in relation to set A when $A \cup B = A$?

10. Consider two finite sets A and B such that $A \cap B = A$.

    (a) Give three examples of sets that satisfy this property.

    (b) In each of the three examples generated in part (a), how does set A relate to set B? Describe this relationship or condition using terms introduced in this section.

    (c) In general, what can one say about set A in relation to set B when $A \cap B = A$?

Paper Assignment 3.1

1. **Summarizing.** Write a short summary of Section 3.1. Among other terms, your summary should include the following, but not necessarily in this order: *set, element, subset, union, intersection*, and *partition*. The explanation of these concepts should not appear as a list but should be connected, such that your summary takes the form of an essay. Where notation is used, follow the format in the text.

Group Activities 3.1

1. Complete Exercise 3.1.4 as a group activity.
2. Complete Exercise 3.1.7 as a group activity.

Further Reading

Halmos, P. R., *Naive Set Theory*, Springer-Verlag, New York, 1974.

3.2 Creating Identification Numbers

Although the set theory concepts presented in the first section of this chapter are not crucial to understanding the process of creating an identification number, they are necessary for investigating functions and permutations, two topics that are motivated through the material presented next.

Preliminary Activity. Suppose that the State of New York decides to change the identification number system that it currently uses to identify drivers. The new method will create a driver's license number using a person's name and birth date. Listed below are the names and birth dates of four drivers and what their driver's license number will be under this new system. Use this information to determine what system New York State will be using to assign driver's license numbers.

Name	Birth Date	License Number
Joseph H. Kirtland	January 1, 1978	KIRTJ0801AA78
Mary M. Cutihee	February 26, 1976	CUTIM1302AZ76
Fred C. Doe	October 30, 1962	DOE*F0310BD62
Elizabeth Rice	July 7, 1942	RICEE* * 07AG42

The reason identifications numbers are created is so that the information they represent can be more easily stored, retrieved, and transmitted. Recall the 12-digit Version A UPC codes discussed in Chapter 2. In a 12-digit UPC a_1-$a_2a_3a_4a_5a_6$-$a_7a_8a_9a_{10}a_{11}$-a_{12}, the first digit a_1, which is the number system character, identifies the type of product (the values are given in Table 2.1). The second set of numbers $a_2a_3a_4a_5a_6$ identifies the manufacturer. The third set of numbers $a_7a_8a_9a_{10}a_{11}$ identifies the product, and the last digit a_{12} is the check digit. Through the UPC, stores can track inventory and identify which types and brands of products are the best (and worst) sellers.

The process of creating an identification number is called *hashing*.

Definition 3.2.1. ***Hashing*** *is the process of taking a certain amount of information and storing that data in an identification number that represents the original information.*

Definition 3.2.2. *A* ***hash*** *or* ***hashing function*** *is the actual process that transforms the information into an identification number.*

Consider the following simple hashing function. It takes an individual and associates with that person the first four letters of the person's last name. If a last name has three or fewer letters in it, each of the remaining spaces is filled with an asterisk, $*$. This hashing function would identify the following last names with the four-character identification numbers shown:

$$\begin{aligned} \text{Clark} &\longrightarrow \text{CLAR} \\ \text{Kirtland} &\longrightarrow \text{KIRT} \\ \text{Thompson} &\longrightarrow \text{THOM} \\ \text{Zia} &\longrightarrow \text{ZIA}* \end{aligned}$$

This hashing function will run into problems if it needs to assign identification numbers to two different people whose last names are Hacker and Hackett. Using the process described above, both individuals would be assigned the same identification number HACK. When this happens, a *collision* occurs. This presents a problem, since we cannot know which individual the identification number HACK refers to. Complications are likely to occur if HACK is used to retrieve information on either Hacker or Hackett. This confusion defeats the purpose of having an identification number system in the first place.

Definition 3.2.3. *A* ***collision*** *occurs when a hashing function assigns the same identification number to two different pieces of information.*

A hashing function is only as good as the number of collisions it avoids. As the number of possible collisions increases, the weaker the hashing function becomes. While it is often impossible to avoid collisions, we want to prevent them as much as possible.

Consider the ISBN hashing process presented in Chapter 2. It takes information about a book (language or country published in, publishing company, specific book) and creates a 12-digit number to identify that book. If two distinct books had the same ISBN, it would defeat the purpose of having ISBNs in the first place. One could not identify specific books with an ISBN.

Now consider the specific hashing function used by the State of Washington and the Province of Manitoba to create driver's license numbers [5]. The hashing function uses the driver's name and date of birth to generate a 12-character identification number $a_1a_2a_3a_4a_5a_6a_7a_8a_9a_{10}a_{11}a_{12}$:

$$\underbrace{\underline{a_1}\ \underline{a_2}\ \underline{a_3}\ \underline{a_4}\ \underline{a_5}}_{1}\ \underbrace{\underline{a_6}}_{2}\ \underbrace{\underline{a_7}}_{3}\ \underbrace{\underline{a_8}\ \underline{a_9}}_{4}\ \underbrace{\underline{a_{10}}}_{5}\ \underbrace{\underline{a_{11}}}_{6}\ \underbrace{\underline{a_{12}}}_{7}$$

1. The first five characters $a_1a_2a_3a_4a_5$ are the first five letters of the individual's surname (last name). If a surname contains four or fewer letters, the asterisk symbol $(*)$ is used to fill in the remaining spaces.
2. The sixth character a_6 is the individual's first initial.
3. The seventh character a_7 is the individual's middle initial. The asterisk $*$ is used when a person has no middle name.
4. The next two characters a_8a_9 are obtained by subtracting the last two digits of the individual's year of birth (YOB) from 100. That is, $100-\text{YOB}$.

TABLE 3.1
Washington/Manitoba Code for Months

Month	Code	Alternative Code	Month	Code	Alternative Code
January	B	S	July	M	4
February	C	T	August	N	5
March	D	U	September	O	6
April	J	1	October	P	7
May	K	2	November	Q	8
June	L	3	December	R	9

5. Character a_{10} is the check digit. How to calculate it will be presented later.
6. The 11th character a_{11} is the month-of-birth code. It assigns a character according to the code presented in Table 3.1. The alternative codes are used to avoid collisions.
7. The last character a_{12} is the code for the day of the month on which the individual was born. The *day-of-birth* code is presented in Table 3.2.

This hashing function would create the following driver's license identification numbers for these three individuals. The tenth position, which contains the check digit, will be left as a_{10} for now. Note that the second person has no middle name, and the third surname has only three letters.

$$\left.\begin{array}{c}\text{Joseph Marshall Pollan}\\ \text{Born: October } 19, 1970\end{array}\right\}\quad \text{POLLAJM30}\underline{a_{10}}\text{PR}$$

$$\left.\begin{array}{c}\text{Tracy Holland}\\ \text{Born: April } 25, 1945\end{array}\right\}\quad \text{HOLLAT} * 55\underline{a_{10}}\text{J5}$$

$$\left.\begin{array}{c}\text{Alexander William Moe}\\ \text{Born: July } 4, 1976\end{array}\right\}\quad \text{MOE} * *\text{AW24}\underline{a_{10}}\text{MD}$$

Now suppose that the identification number created above for Joseph Marshall Pollan is already in the system. In creating an identification number for Joseph's twin brother John Michael, we would use the alternative month code to avoid a collision (assuming

TABLE 3.2
Washington/Manitoba Code for Days

1 - A	6 - F	11 - J	16 - W	21 - 1	26 - 6	31 - U
2 - B	7 - G	12 - K	17 - P	22 - 2	27 - 7	
3 - C	8 - H	13 - L	18 - Q	23 - 3	28 - 8	
4 - D	9 - Z	14 - M	19 - R	24 - 4	29 - 9	
5 - E	10 - S	15 - N	20 - 0	25 - 5	30 - T	

TABLE 3.3
Washington/Manitoba Character-to-Digit Assignments

* - 0	D - 4	H - 8	L - 3	P - 7	T - 3	X - 7
A - 1	E - 5	I - 9	M - 4	Q - 8	U - 4	Y - 8
B - 2	F - 6	J - 1	N - 5	R - 9	V - 5	Z - 9
C - 3	G - 7	K - 2	O - 6	S - 2	W - 6	

that Joseph and John were born on the same day).

John Michael Pollan
Born: October 19, 1970 } POLLAJM30$\underline{a_{10}}$7R

Although this hashing function works well, it cannot avoid collisions. Currently, both Joseph Marshall Pollan, born on October 19, 1970, and John Michael Pollan, born on October 19, 1970, are in the system. Suppose a new driver, Jessica Mosher Pollard, also born on October 19, 1970, were to be entered into the system. A collision could not be avoided unless the hashing function used to create these driver's license numbers were altered.

We have not yet addressed the check digit scheme [5]. The preceding driver's license numbers were created with no check digits.

Definition 3.2.4. *Consider the 12-character Washington State and Manitoba Province driver's license identification number* $a_1a_2a_3a_4a_5a_6a_7a_7a_9a_{10}a_{11}a_{12}$. *The check digit* a_{10} *is created by first assigning digits to the non-digit characters in the identification number, using Table 3.3, and then performing the following calculation:*

$$a_{10} = | \, a_1 - a_2 + a_3 - a_4 + a_5 - a_6 + a_7 - a_8 + a_9 - a_{11} + a_{12} \, | \pmod{10}$$

The driver's license number for Joseph Marshall Pollan is POLLAJM30a_{10}PR, where a_{10} is the check digit. To calculate the check digit, all non-digit characters must first be converted into digits using Table 3.3, and then the check digit a_{10} can be computed. First, the conversion is performed, as illustrated in Figure 3.1.

a_1	a_2	a_3	a_4	a_5	a_6	a_7	a_8	a_9	a_{10}	a_{11}	a_{12}
P	O	L	L	A	J	M	3	0	a_{10}	P	R
↓	↓	↓	↓	↓	↓	↓	↓	↓	↓	↓	↓
7	6	3	3	1	1	4	3	0	a_{10}	7	9

FIGURE 3.1
Converting Characters to Digits for License POLLAJM30a_{10}PR

Then, the calculation:

$$\begin{aligned} a_{10} &= |\, a_1 - a_2 + a_3 - a_4 + a_5 - a_6 + a_7 - a_8 + a_9 - a_{11} + a_{12} \,| \pmod{10} \\ &= |\, 7 - 6 + 3 - 3 + 1 - 1 + 4 - 3 + 0 - 7 + 9 \,| \pmod{10} \\ &= |\, 4 \,| \pmod{10} = 4 \pmod{10} = 4. \end{aligned}$$

As a result, the full driver's license number for Joseph Marshall Pollan would be POLLAJM304PR. Applying this process to the driver's license numbers created above, the full number for Tracy Holland would be HOLLAT∗554J5; Alexander William Moe's would be MOE∗∗AW240MD; and John Michael Pollan's would be POLLAJM3047R.

The question now is: How good is this scheme?

The goal is for it to catch all the errors listed in Table 1.2. However, at the very least, a scheme should catch every single-digit error and every transposition-of-adjacent-digits error. Consider the valid driver's license identification number POLLAJM304PR that was just created. Listed below are two different single-digit errors:

Correct Number:	POLL**A**JM304PR	POLLAJM304**P**R
	↓	↓
Incorrect Number:	POLL**R**JM304PR	POLLAJM304**G**R

In the first case, this scheme catches the error. First, Table 3.3 is used to generate the sequence of 12 digits to be used in the check digit calculation: POLL**R**JM304PR = 763391430479. This leads to the calculation

$$\begin{aligned} |\, 7 - 6 + 3 - 3 + 9 - 1 + 4 - 3 + 0 - 7 + 9 \,| \pmod{10} &= |\, 12 \,| \pmod{10} \\ &= 12 \pmod{10} = 2. \end{aligned}$$

Since the check digit is 4, the error is caught.

In the second case, this scheme does not catch the error. First, Table 3.3 is used to generate the sequence of 12 digits for the check digit calculation: POLLAJM304**GR** = 763311430479. This leads to the calculation

$$\begin{aligned} |\, 7 - 6 + 3 - 3 + 1 - 1 + 4 - 3 + 0 - 7 + 9 \,| \pmod{10} &= |\, 4 \,| \pmod{10} \\ &= 4 \pmod{10} = 4. \end{aligned}$$

Since the check digit is 4, the error is not caught.

This scheme catches some, but not all, single-digit errors. By working with a few more examples, it can be readily seen that this scheme does not catch all transposition-of-adjacent-digits errors either.

Exercises 3.2

1. Consider the following Washington State driver's license numbers. Determine which number is valid and which is invalid.

   SMITHML370UZ HART∗JM243MD

2. Create a Washington State driver's license number, complete with check digit, for Janet Mary Huntington, born on February 17, 1981.

3. The federal government wants to set up an identification number system to keep track of all students who have received a government-sponsored loan. The identification number, denoted $a_1a_2a_3a_4a_5a_6a_7a_8a_9a_{10}a_{11}a_{12}a_{13}a_{14}a_{15}$, is to be 15 characters long and will encode a student's name, college being attended, year of graduation, and loan amount. The hashing function works as follows:

   (a) The first four characters $a_1a_2a_3a_4$ are the first four letters of the student's last name. For a surname with only one to three letters, add enough pound signs (#) to make a total of four characters. For example, MOE becomes MOE#.

   (b) The next character a_5 is the first letter of the student's first name.

   (c) The next three characters $a_6a_7a_8$ are the initials of the college. If there are only two, add a # at the end. With four or more initials, use only the first three. For example, Marist College becomes MC# and the University of California at Los Angeles becomes UCL.

   (d) The next two characters a_9a_{10} are the last two digits of the year the student will or did graduate. For example, 91 denotes graduation from college in the year 1991.

   (e) The last five digits $a_{11}a_{12}a_{13}a_{14}a_{15}$ are the amount of the loan. Add enough zeros in front of the number to make it five digits long. For example, if the loan is for $4500, write 04500.

   For example, the student Joe Kirtland, who attended Syracuse University and graduated in the year 1999 with a loan of $350, would have the following identification number: KIRTJSU#9900350.

   Identify two different situations where a collision could occur when using this hashing function. Explain each type of collision and give an example of each one.

4. Listed below are five individuals, their dates of birth, cities of residence, and an identification number associated with each person. Use the personal information given to determine the hashing function used to create the identification numbers.

Name	Information	Identification Number
Bob Harvey Smith	March 26, 1960 New York City	SMITH#02H10C6NE
Bruce Hawley Smith	March 26, 1960 New York City	SMITH#02H10*O*6NE
John Tracy Smith	September 1, 1973 Buffalo	SMITH#10T23IABU
Joseph Thomas	January 1, 1941 Albany	THOMAS10#91AAAL
Sally Edison Paulison	December 29, 1993 Syracuse	PAULIS19E43L9SY

Paper Assignment 3.2

1. **Process Description and Argumentation.** You have been hired by the New York State Department of Motor Vehicles to develop a new driver's license identification number system for the state. The new identification numbers must encode a person's first and last name, date of birth, and county resided in. In addition, the identification number must be only ten characters long, including the check digit. Consequently, besides creating a hashing function, you need to develop a check digit scheme for this identification number system.

   Write a paper, in the form of a report, presenting the new driver's license identification number system (with check digit scheme) to the head of the Department of Motor Vehicles. Make a strong case for the adoption of your system by doing the following:

   (a) Give a step-by-step description, with sufficient examples, of how a driver's personal information would be used.

   (b) Include a detailed description of the hashing function and the check digit scheme.

   (c) Argue that your system avoids most, if not all, types of collisions, using examples as needed.

Group Activities 3.2

1. Complete Exercise 3.2.4 as a group activity.
2. A driver's license number from the State of Utah consists of nine digits, where $a_1a_2a_3a_4a_5a_6a_7a_8$ identifies the driver and a_9 is the check digit. Using the following list of valid driver's license numbers, determine the check digit scheme employed by Utah. (HINT: It is a "mod 10" scheme that involves addition and multiplication.)

122351606	122351711	122352610	122351630
122351614	122351517	122353616	122351313
122351622	122351410	122354612	122355618

Further Reading

Gallian, J. A., Assigning Driver's License Numbers, *Mathematics Magazine*, 64(1), 1991, 13–22.

Gallian, J. A., Breaking the Missouri License Code, *UMAP Journal*, 13(1), 1992, 36–41.

3.3 Functions

Hashing functions establish a correspondence between a list of individual pieces of information and a list of identifying numbers. This kind of correspondence will be investigated further in this section through a more general study of functions. Sets are central to this examination. Special types of functions, called permutations, will be studied after our investigation of functions is complete. This is important because permutations help create check digit schemes much more sophisticated than the ones discussed so far.

Preliminary Activity. In Chapter 2, the method used to assign a UPC to a product was described. In the previous section, the method used by the State of Washington and the Province of Manitoba, Canada, to assign driver's license numbers was presented. Both are examples of hashing, where information (product information for the UPC and personal information for the license) is associated with or corresponds to an identification number.

Look in a print or electronic media source and find an example of a correspondence between two sets. The hashing method does not need to be made explicit. Simply find an example where each item in one set corresponds to an item in another set. Write a few sentences that describe the correspondence you found.

Functions are used extensively in everyday life. Consider the following two data tables. Table 3.4 gives the US national public debt, in billions of dollars, for the years 1978 through 1997. Table 3.5 presents the major exports of several countries in Europe.

TABLE 3.4
US National Public Debt in Billions of Dollars

Year	Debt	Year	Debt	Year	Debt
1978	789.2	1985	1945.9	1992	4064.6
1979	845.1	1986	2125.3	1993	4411.4
1980	930.2	1987	2350.2	1994	4692.7
1981	1028.7	1988	2602.3	1995	4973.9
1982	1197.1	1989	2857.4	1996	5224.8
1983	1410.7	1990	3233.3	1997	5413.1
1984	1662.9	1991	3665.3		

TABLE 3.5
Major Exports of Several European Countries

Country	Major Exports
Austria	Machinery and transportation equipment, manufactured goods, paper and paper products, chemicals
France	Automobiles, electrical machinery, metal products
Germany	Motor vehicles and parts, industrial machinery and equipment, electrical products, iron and steel products
Italy	Machinery and manufactured products, textiles, clothing, shoes, chemicals, transport equipment
Portugal	Clothing, footwear, paper and paper products, ships and boats, cork and other wood products
Spain	Transport equipment, agricultural produce, beverages
Switzerland	Clocks, watches, precision instruments, turbines, generators, chemicals, cheese, chocolate

Do you notice any similarities between the Tables 3.4 and 3.5? Do you notice any differences? Are sets involved? If so, what are they? Each table presents two different categories, or sets, and the role that each set plays is different. How so? There is some sort of correspondence between the two sets in each example. Is the correspondence in one table different from the correspondence in the other?

In the public debt table, two sets are being used. The first is the set of years A between 1978 and 1997, or $A = \{1978, 1979, 1980, \ldots, 1997\}$. The second is the set of numbers $B = \{789.2, 2602.3, 845.1, 2857.4, 930.2, 3233.3, 1028.7, 3665.3, 1197.1, 4064.6, 1410.7, 4411.4, 1662.9, 4692.7, 1945.9, 4973.9, 2125.3, 5224.8, 2350.2, 5413.1\}$ in which each number represents an amount of money in billions of dollars. Each year corresponds to exactly one dollar amount (representing the public debt that year).

In the major export table, there are also two sets involved. The first set C contains the European countries whose exports are listed: $C = \{$Austria, France, Germany, Italy, Portugal, Spain, Switzerland$\}$. The second set D is the collection of exports, where $D = \{$machinery and transportation equipment, manufactured goods, paper and paper products, chemicals, automobiles, electrical machinery, metal products, motor vehicles and parts, industrial machinery and equipment, electrical products, iron and steel products, machinery and manufactured products, textiles, clothing, shoes, chemicals, transport equipment, clothing, footwear, paper and paper products, ships and boats, cork and other wood products, agricultural produce, beverages, clocks, watches, precision instruments, turbines, generators, cheese, chocolate$\}$.

The public debt table is an example of a *function*, while the major export table is not.

Definition 3.3.1. *A **function** f is a relation or correspondence between two sets A and B where each element $a \in A$ corresponds to at most one element $b \in B$.*

The letter f is not the only one used to denote a function. The letters g, h, k, and so on are also used. When f is a function between sets A and B, it is denoted

$$f : A \to B.$$

When an element $a \in A$ corresponds to an element $b \in B$, we write

$$f(a) = b.$$

The hashing functions described in the previous section are all examples of functions. Each one establishes a correspondence between a set of items or individuals and a set of identification numbers. Furthermore, each item or individual corresponds to at most one identification number. For the Washington State driver's license system, the hashing function establishes a correspondence between the set of drivers in the state and a set of identification numbers, where each driver is assigned a unique (only one) license identification number.

Let f denote the US national public debt function presented in Table 3.4. The fact that in 1985 the US public debt was 1945.9 billion dollars can be represented by $f(1985) = 1945.9$. The notation $f(1997) = 5413.1$ indicates that the public debt in 1997 was 5413.1 billion dollars.

The major exports table is not an example of a function, since one element from set C, say Austria, does not correspond to at most one element from set D. In fact, Austria corresponds to several elements, namely, machinery and transportation equipment, manufactured goods, paper and paper products, and chemicals.

Another way to present the correspondence established in a function $f : A \to B$ is to list the elements of A and B vertically and side by side, and then to draw an arrow from each element $a \in A$ to the element $b \in B$ $(a \to b)$ that it is related to. For instance,

$$\begin{array}{ccc} \underline{A} & \xrightarrow{f} & \underline{B} \\ \vdots & & \vdots \\ a & \longrightarrow & b \\ \vdots & & \vdots \end{array}$$

Given a function $f : A \to B$, each element $a \in A$ must correspond to at most one element $b \in B$. Consider the function $f : A \to B$, with $A = \{2, 6, 9\}$ and $B = \{1, 3, 7\}$. Define the correspondence by $f(2) = 7$, $f(6) = 1$, and $f(9) = 3$. This can be represented in a diagram as follows:

$$\begin{array}{ccc} \underline{A} & \xrightarrow{f} & \underline{B} \\ 2 & & 1 \\ 6 & & 3 \\ 9 & & 7 \end{array} \qquad (2 \to 7,\ 6 \to 1,\ 9 \to 3)$$

The correspondence given in Table 3.5 (major exports) was not a function, because each single element in set C corresponded to more than one element from set D. Example 3.3.2 presents a similar case.

Example 3.3.2. *The correspondence g between the set $A = \{1, 2, 3, 4\}$ and the set $B = \{2, 4, 6, 8, 10\}$ is not a function, because $g(1) = 2$ and $g(1) = 4$.*

$$\begin{array}{ccc} \underline{A} & \xrightarrow{g} & \underline{B} \\ 1 & & 2 \\ 2 & & 4 \\ 3 & & 6 \\ 4 & & 8 \\ & & 10 \end{array} \qquad (1 \to 2,\ 1 \to 4,\ 2 \to 6,\ 3 \to 8,\ 4 \to 10)$$

However, given a function f, each element $b \in B$ can have many elements $a \in A$ corresponding to it, as Example 3.3.3 illustrates.

Example 3.3.3. *The correspondence f between the set $A = \{1, 2, 3, 4, 5\}$ and the set $B = \{2, 4, 6\}$ is a function, even though the element $2 \in B$ has three elements from A corresponding to it: $f(1) = 2$, $f(2) = 2$, and $f(3) = 2$.*

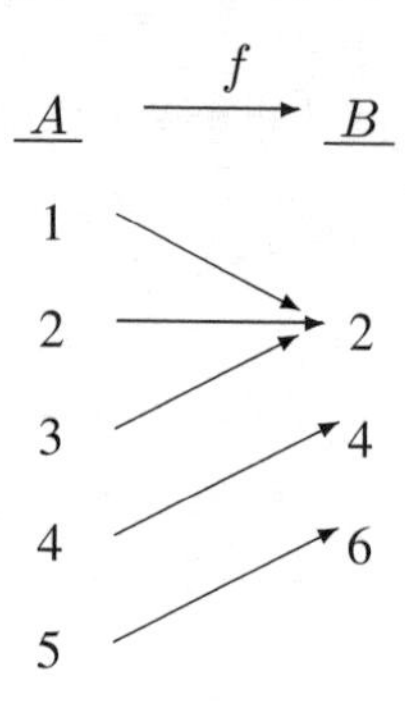

A function can also establish a correspondence between elements of the same set. This is demonstrated in Example 3.3.4.

Example 3.3.4. *The function $f : A \to A$ illustrates a correspondence from the set $A = \{0, 1, 2, 3, 4, 5\}$ to itself.*

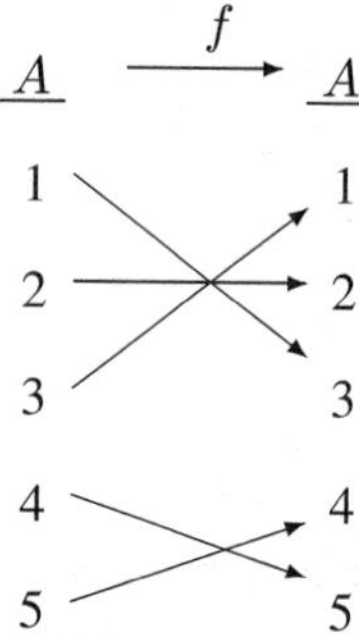

In addition, given a function from a set A to a set B, not every element in A must correspond to an element in B. Furthermore, not every element in B needs to have an element in A corresponding to it. Example 3.3.5 depicts both cases.

Example 3.3.5. *In the function f from set $A = \{1, 2, 3, 4, 5\}$ to set $B = \{2, 7, 9\}$, both 2 and 4 from A correspond to nothing in B. In the function g from the set $A = \{1, 2, 3, 4\}$ to the set $B = \{2, 5, 7, 8, 9\}$, the element $8 \in B$ has no element from A corresponding to it.*

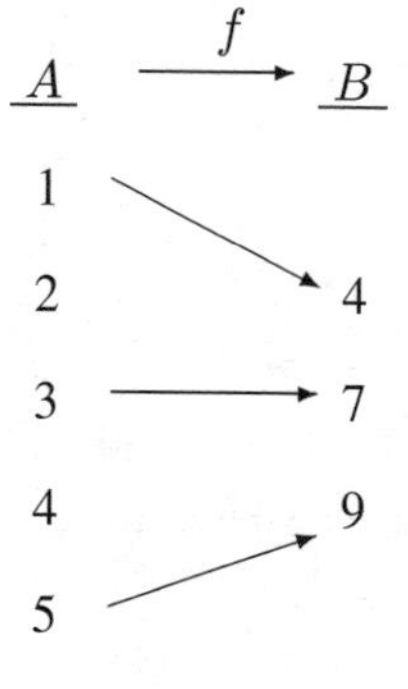

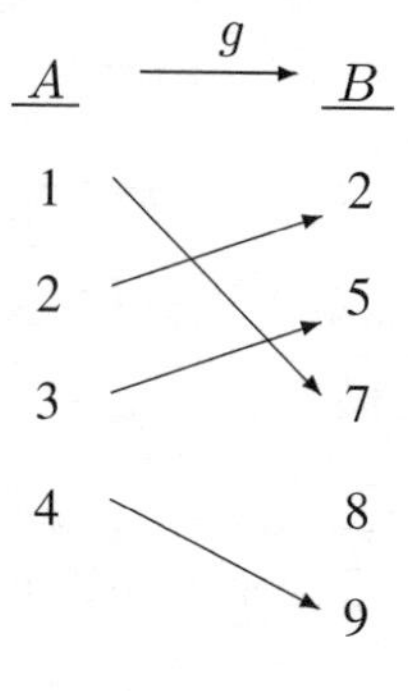

The functions f and g from Example 3.3.5 motivate the following definitions.

Definition 3.3.6. *Let f be a function from a set A to a set B ($f : A \to B$). The collection of elements $a \in A$ that correspond to some element $b \in B$ is called the* **domain** *of f and is denoted dom(f).*

Definition 3.3.7. *Let f be a function from a set A to a set B ($f : A \to B$). The collection of elements $b \in B$ for which there is an element $a \in A$ such that $f(a) = b$ is called the* ***range*** *of f and is denoted ran(f).*

For the function f presented in Example 3.3.5, $\text{dom}(f) = \{1, 3, 5\}$ and $\text{ran}(f) = \{4, 7, 9\} = B$. For the function g from Example 3.3.5, $\text{dom}(g) = \{1, 2, 3, 4\} = A$ and $\text{ran}(g) = \{2, 5, 7, 9\}$. Functions from A to B exist where the domain is all of set A and the range is all of set B.

Besides having different domains and ranges, other characteristics differentiate functions from one another. To motivate one of them, recall the work done in the previous section on hashing functions and the Washington State driver's license identification numbers. The hashing function created a correspondence between the drivers in the state and a set of identification numbers. This correspondence was clearly a function, because each driver was assigned his or her own number. However, one problem that was unavoidable was that two different drivers could be assigned (could correspond to) the same driver's license number.

Recall the situation that occurred in assigning driver's license numbers to Joseph Marshall Pollan, born on October 19, 1970, John Michael Pollan, born on October 19, 1970, and Jessica Mosher Pollard, born on October 19, 1970. Joseph M. Pollan was assigned the number POLLAJM304PR. Given that John has the same initials and was born the same day and year as Joseph, the alternative month code was used to assign his number of POLLAJM3047R. Now, Jessica Mosher Pollard, also born on October 19, 1970, still needs a number. Since the alternative month code has already been used, a collision will occur. Unless changes are made to the hashing function, she must be assigned the driver's license number POLLAJM304PR or POLLAJM3047R. In either case, two people will get the same number and a collision results. Collisions and another concept are explored next.

Example 3.3.8. *Compare the following three functions g, h, and k. Besides having different domains and ranges, fundamental properties separate these three functions.*

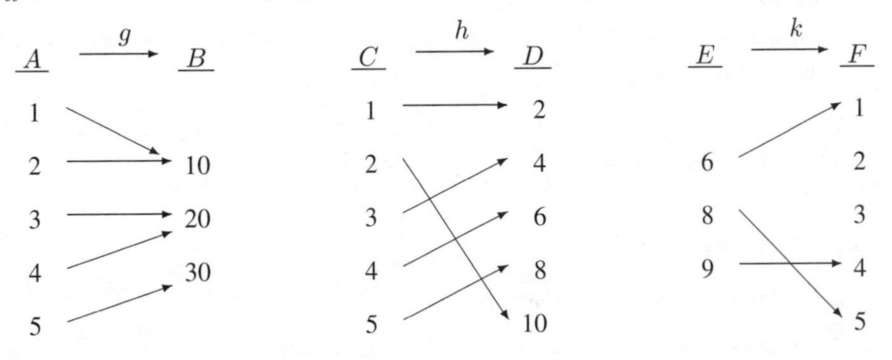

Not all functions are created equal. Some have special properties.

Definition 3.3.9. *A function* $f : A \to B$ *is* ***one-to-one*** *if for each pair of elements* a_1 *and* a_2 *in* A, *where* $a_1 \neq a_2$, $f(a_1) \neq f(a_2)$.

Definition 3.3.10. *A function* $f : A \to B$ *is* ***onto*** *if for each element* $b \in B$, *there is an element* $a \in A$ *such that* $f(a) = b$.

Look at the functions that were given in Example 3.3.8. The function g is not one-to-one, because $g(1) = g(2) = 10$. It is onto, because each element in B has an element in A corresponding to it. The function h is both one-to-one and onto. The function k is one-to-one, but it is not onto, as no element from the set A corresponds to the element 2 or the element 3 in set B.

Exercises 3.3

1. In this section, many examples of functions were given. In particular, we studied correspondences motivated by real data (e.g., the US national public debt). Look in a newspaper, magazine, or other source and find two examples of a correspondence that results in functions. For each example you find, indicate what correspondence the function is establishing, whether the function is one-to-one, onto, or both, and what the domain and range of the function are.
2. Consider these two sets.

$$A = \{1,2,3,4,5,6,7,8\} \qquad B = \{-1,-2,-3,-4,-5,-6,-7,-8\}$$

   (a) Create an arbitrary function $f : A \to B$.

   (b) Can a function $f : A \to B$ be onto? If so, create one; if not, explain why.

   (c) Can a function $f : A \to B$ have $\text{dom}(f) = A$? If so, create one; if not, explain why.

   (d) Can a function $f : A \to B$ have $\text{dom}(f) = A$ and be one-to-one, but not onto? If so, create one; if not, explain why.

3. (a) Let the set $A = \{0,2,4,6,8,10\}$ and the set $B = \{12,14,16,18,20\}$. Explain and demonstrate why a function $f : A \to B$ where $\text{dom}(f) = A$ cannot be one-to-one.

   (b) Let the set $A = \{1,2,3,4,5\}$ and the set $B = \{1,2,3,4,5,6,7\}$. Explain and demonstrate why a function $f : A \to B$ cannot be onto.

4. Let $f : A \to B$ be a function from set A to set B, where sets A and B are finite and $\text{dom}(f) = A$. What can you say about the $\mid A \mid$ in comparison to $\mid B \mid$ if

   (a) f is one-to-one?

   (b) f is onto?

   (c) f is both one-to-one and onto?

   Be sure to explain how you came up with your answer. If it helps, give examples that support each answer. Recall the work you did in Exercises (2) and (3).

5. Determine which of the following are functions. Make sure you clearly indicate whether each one is a function or not and how you arrived at your answer.

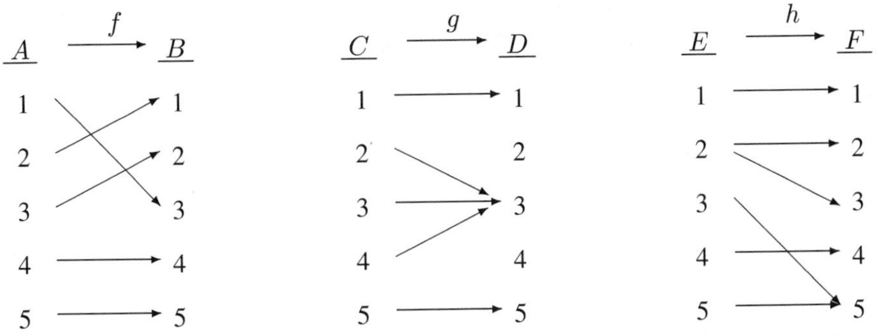

$A \xrightarrow{f} B$ $C \xrightarrow{g} D$ $E \xrightarrow{h} F$

1 1   1 1   1 1
2 2   2 2   2 2
3 3   3 3   3 3
4 4   4 4   4 4
5 5   5 5   5 5

6. Consider the following three functions. For each function, determine whether it is one-to-one, onto, or both. Be sure to support your answers.

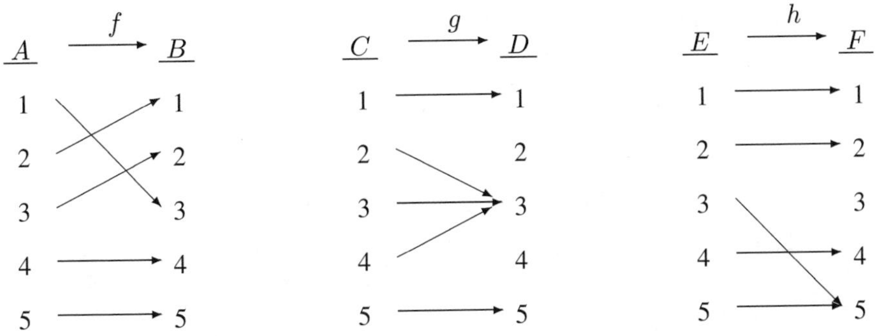

$A \xrightarrow{f} B$ $C \xrightarrow{g} D$ $E \xrightarrow{h} F$

1 1   1 1   1 1
2 2   2 2   2 2
3 3   3 3   3 3
4 4   4 4   4 4
5 5   5 5   5 5

7. Consider the following functions $f : A \to B$ and $g : B \to C$, where $A = \{1, 2, 3, 4, 5\}$, $B = \{-1, -2, -3, -4, -5\}$, and $C = \{2, 4, 6, 8, 10\}$. There is a way to "combine" the functions f and g to create a new function $h : A \to C$. How might this be done? Present all the details of your work.

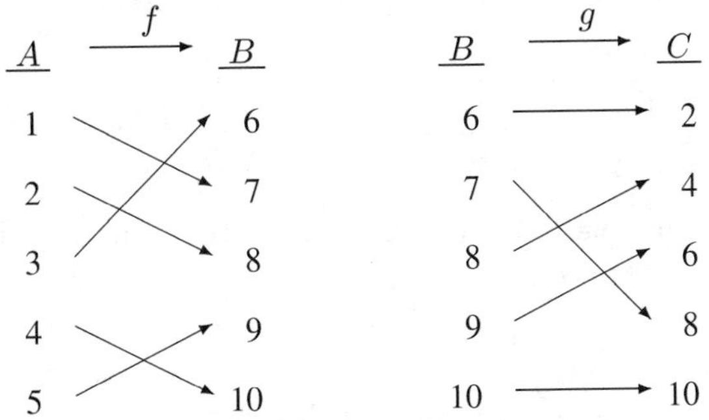

$A \xrightarrow{f} B$ $B \xrightarrow{g} C$

1 6   6 2
2 7   7 4
3 8   8 6
4 9   9 8
5 10   10 10

Paper Assignments 3.3

1. **Defining.** In this section, many examples of correspondences were given. In particular, we studied correspondences motivated by real data (the US national public debt

and exports of selected European countries). Look in a newspaper, magazine, or other source and find two examples of correspondences, one that results in a function and one that does not. In a short paper, present both of the correspondences and explain why one results in a function and the other does not.

2. **Defining.** In this section, many examples of functions were given. In particular, we studied a function motivated by real data (e.g., the US national public debt). Look in a newspaper, magazine, or other source and find two examples of correspondences that result in functions. In a short paper, present each function, indicate what correspondence each is establishing, whether it is one-to-one or onto (or both), and what its domain and range are.

Group Activities 3.3

1. (a) Let $f : A \to B$ be a function from set A to set B, where sets A and B are finite and $\text{dom}(f) = A$. Each member of the group will create three functions $f : A \to B$ where

   i. the first function has $\mid A \mid$ less than $\mid B \mid$,

   ii. the second function has $\mid A \mid$ equal to $\mid B \mid$, and

   iii. the third function has $\mid A \mid$ greater than $\mid B \mid$.

   (b) Compare your examples within the group. Based on these examples, what can you say about the $\mid A \mid$ in comparison to $\mid B \mid$ in the following cases?

   i. f is one-to-one.

   ii. f is onto.

   iii. f is both one-to-one and onto.

2. In this section, you investigated one-to-one and onto functions $f : A \to B$ with finite sets A and B. There are also such functions f when A and B are infinite sets.

   Consider the following two pairs of infinite sets:

   (a) $A = \{1, 2, 3, 4, 5, \dots\}$ and $B = \{-1, -2, -3, -4, -5, \dots\}$;

   (b) $A = \{1, 2, 3, 4, 5, \dots\}$ and $B = \{2, 4, 6, 8, 10, \dots\}$.

   For each pair of infinite sets A and B, create three functions $f : A \to B$ where the first is one-to-one but not onto, the second is onto but not one-to-one, and the third is both one-to-one and onto. Since the sets involved are infinite, it is hard to present these functions through the use of a diagram. Functions that involve infinite sets are usually presented by using a formula that indicates how the correspondence is created.

Further Reading

Bernard K. J., and Wellenzohn, H. J., *Foundations of Mathematics*, H&H Publishing Company, Clearwater, FL, 1997.

3.4 Permutations

Preliminary Activity. Consider the sets $A = \{0, 1, 2\}$ and $B = \{3, 4, 5\}$.

1. There are 27 different functions $f : A \to B$ where $\text{dom}(f) = A$. Find and record as many as you can.
2. Develop a "shorthand" method to represent each of the functions found in part (1).
3. Take the collection of functions found in part (1) and subdivide it into at least three separate categories. Each category should contain functions that share a similar property. Write a few sentences that describe the common characteristics of the functions listed in each category.

As described earlier, each driver in Washington State is assigned a 12-character driver's license number. When driving and criminal records are computerized, this license number is used to retrieve such information. Since each driver corresponds to only one 12-digit number, this correspondence determines a function. Table 3.6 provides some examples.

TABLE 3.6
Driver's License Assignments

Joseph Marshall Pollan	$\to$	POLLAJM304PR
John Michael Pollan	$\to$	POLLAJM3047R
Tracy Holland	$\to$	HOLLAT∗554J5
Alexander William Moe	$\to$	MOE∗∗AW240MD

A collision would occur if two different drivers were assigned the same identification number. As listed in Table 3.6, Joseph Marshall Pollan was assigned the number POLLAJM304PR, and John Michael Pollan was assigned the number POLLAJM3047R. Since John has the same last name, first initial, middle initial, and birth date as Joseph, the alternative month code was used to determine his driver's license number.

Now, suppose that Jessica Mosher Pollard, born on October 19, 1970, needs to be assigned a number. The alternative month code has already been used, so a collision will occur unless the hashing function is somehow altered. Without such an alteration, her driver's license number will be either POLLAJM304PR or POLLAJM3047R. Either way, two people will be assigned the same number and a collision will occur. It is because the function is not one-to-one that problems result. When Jessica's number is typed into a computer, either Joseph's or John's records may be retrieved. Consequently, having no collisions (a one-to-one function) is very desirable. This function should be onto, as well. Otherwise, extra numbers will be floating around in the system without any drivers associated with them. This could cause data entry and retrieval problems.

One-to-one and onto functions, called *permutations*, are a special class of functions that merit careful study. In the following pages, various properties of permutations will be established. One important application of permutations is in the area of check digit schemes. Check digit schemes that use a permutation are often more sophisticated and catch more transmission errors than others that do not.

Definition 3.4.1. *A function $f : A \to B$ is a* ***permutation*** *if it is both one-to-one and onto.*

Consider the following correspondences g, h, and k between the set $A = \{0, 1, 2, 3, 4, 5, 6, 7\}$ and the set $B = \{8, 9, 10, 11, 12, 13, 14, 15\}$.

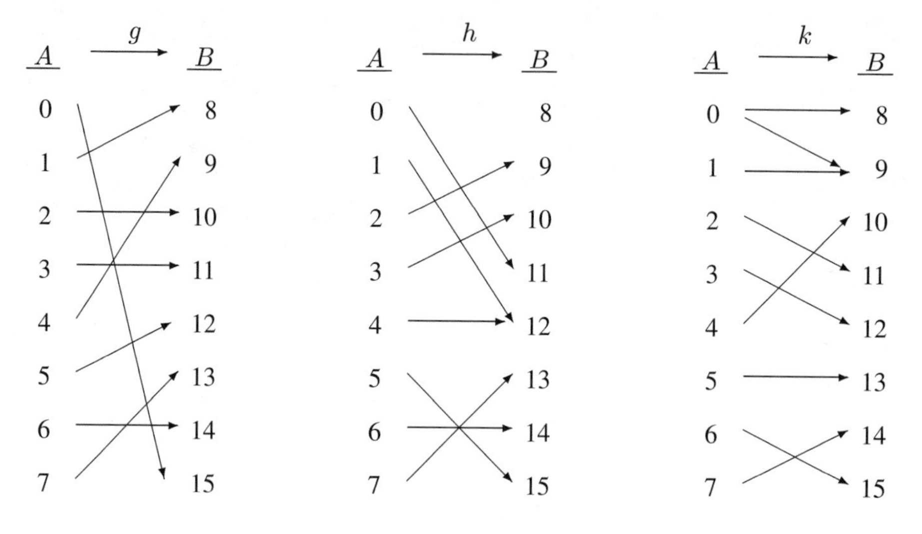

- The function g is a permutation since it is both a one-to-one and onto.
- While h is a function, it is not a permutation. It is not one-to-one, since both $h(1) = 12$ and $h(4) = 12$. It is also not onto, as there is no element in A that corresponds to the element 8 in B.
- The correspondence established in k does not result in a permutation, since k is not even a function: $k(0) = 8$ and $k(0) = 9$.

There are also permutations $f : A \to B$ where both A and B are infinite sets. Consider the set $A = \{0, 1, 2, 3, \dots\}$, the set of whole numbers, and $B = \{0, 2, 4, 6, \dots\}$, the set of non-negative even integers. The function f that sends each element $a \in A$ to $f(a) = 2a = b \in B$ is a permutation.

$$\begin{array}{ccc} A & \xrightarrow{f} & B \\ 0 & \longrightarrow & 0 \\ 1 & \longrightarrow & 2 \\ 2 & \longrightarrow & 4 \\ 3 & \longrightarrow & 6 \\ \vdots & & \vdots \end{array}$$

This function f is one-to-one. If a_1, $a_2 \in A$ and $f(a_1) = f(a_2)$, then

$$\begin{aligned} f(a_1) &= f(a_2) \\ 2a_1 &= 2a_2 \qquad \text{(divide both sides by 2)} \\ a_1 &= a_2. \end{aligned}$$

To show that f is onto, let $b \in B$. Since it is in the set B of positive even numbers, b is a specific even number. Thus b is divisible by 2 and $b = 2x$, for some non-negative number x. Consequently, $f(x) = 2x = b$. Thus there is an element $a \in A$ $(a = x)$, such that $f(a) = b$.

A special type of permutation is a function $f : A \to A$ (a function f from a set A to itself). Consider the set $A = \{0\}$. There is only one permutation $f : A \to A$.

$$A \xrightarrow{f_1} A \qquad 0 \to 0$$

If $A = \{0, 1\}$, there are two permutations $g : A \to A$.

$$A \xrightarrow{g_1} A \qquad \begin{array}{c} 0 \to 0 \\ 1 \to 1 \end{array} \qquad\qquad A \xrightarrow{g_2} A \qquad \begin{array}{c} 0 \to 1 \\ 1 \to 0 \end{array}$$

If $A = \{0, 1, 2\}$, there are six permutations $h : A \to A$.

$$A \xrightarrow{h_1} A \qquad \begin{array}{c} 0 \to 0 \\ 1 \to 1 \\ 2 \to 2 \end{array} \qquad\qquad A \xrightarrow{h_2} A \qquad \begin{array}{c} 0 \to 1 \\ 1 \to 2 \\ 2 \to 0 \end{array} \qquad\qquad A \xrightarrow{h_3} A \qquad \begin{array}{c} 0 \to 2 \\ 1 \to 0 \\ 2 \to 1 \end{array}$$

$$A \xrightarrow{h_4} A \qquad \begin{array}{c} 0 \to 1 \\ 1 \to 0 \\ 2 \to 2 \end{array} \qquad\qquad A \xrightarrow{h_5} A \qquad \begin{array}{c} 0 \to 2 \\ 1 \to 1 \\ 2 \to 0 \end{array} \qquad\qquad A \xrightarrow{h_6} A \qquad \begin{array}{c} 0 \to 0 \\ 1 \to 2 \\ 2 \to 1 \end{array}$$

Drawing these diagrams to present functions gets to be a bit much, therefore a special method has been designed to do so when the function from a finite set A to A is a permutation. Let $f : A \to A$ be a permutation of the set $A = \{a_1, a_2, a_3, \ldots, a_{n-1}, a_n\}$.

Then f can be denoted

$$\begin{pmatrix} a_1 & a_2 & a_3 & \dots & a_{n-1} & a_n \\ f(a_1) & f(a_2) & f(a_3) & \dots & f(a_{n-1}) & f(a_n) \end{pmatrix}.$$

Example 3.4.2. *We use this permutation notation as follows.*

1. The collection of permutations of the set $A = \{0\}$ is denoted

$$\left\{ f_1 = \begin{pmatrix} 0 \\ 0 \end{pmatrix} \right\}.$$

2. The collection or set of permutations of the set $A = \{0, 1\}$ is denoted

$$\left\{ g_1 = \begin{pmatrix} 0 & 1 \\ 0 & 1 \end{pmatrix}, g_2 = \begin{pmatrix} 0 & 1 \\ 1 & 0 \end{pmatrix} \right\}.$$

3. The collection or set of permutations of the set $A = \{0, 1, 2\}$ is denoted

$$\left\{ h_1 = \begin{pmatrix} 0 & 1 & 2 \\ 0 & 1 & 2 \end{pmatrix}, h_2 = \begin{pmatrix} 0 & 1 & 2 \\ 1 & 2 & 0 \end{pmatrix}, h_3 = \begin{pmatrix} 0 & 1 & 2 \\ 2 & 0 & 1 \end{pmatrix}, \right.$$
$$\left. h_4 = \begin{pmatrix} 0 & 1 & 2 \\ 1 & 0 & 2 \end{pmatrix}, h_5 = \begin{pmatrix} 0 & 1 & 2 \\ 2 & 1 & 0 \end{pmatrix}, h_6 = \begin{pmatrix} 0 & 1 & 2 \\ 0 & 2 & 1 \end{pmatrix} \right\}.$$

These collections do not stop with the permutations of a set of three elements. Given any finite set, all the permutations of that set can be constructed. This motivates the following definition.

Definition 3.4.3. *The collection of all the permutations of a set A of order n is called the* ***symmetric group on n elements*** *, denoted S_A. When the set $A = \{0, 1, 2, 3, \dots, n-1\}$, the symmetric group is denoted S_n.*

Remark. The symmetric group S_A is the collection or set of permutations of a set A. Thus an element of S_A is a function, not a number. Also note that S_n is the set of permutations on a set $A = \{0, 1, 2, 3, \dots, n-1\}$ of n elements. It seems unnatural to start this set with 0 and end with $n-1$, but there is an important reason. In this context, 0 is the *first* element of A, 1 is the *second* element of A, 2 is the *third* element of A, continuing until $n-1$, which is the nth element of A. When it comes to identification numbers, S_{10} plays an important role, since it is the set of permutations of the set $\{0, 1, 2, 3, 4, 5, 6, 7, 8, 9\}$, which is the set of all the digits that occur in identification numbers.

Example 3.4.4.

1. The set S_A of all permutations of the set $A = \{2, 5, 9\}$ is

$$S_A = \left\{ \begin{pmatrix} 2 & 5 & 9 \\ 2 & 5 & 9 \end{pmatrix}, \begin{pmatrix} 2 & 5 & 9 \\ 5 & 9 & 2 \end{pmatrix}, \begin{pmatrix} 2 & 5 & 9 \\ 9 & 2 & 5 \end{pmatrix}, \begin{pmatrix} 2 & 5 & 9 \\ 5 & 2 & 9 \end{pmatrix}, \begin{pmatrix} 2 & 5 & 9 \\ 2 & 9 & 5 \end{pmatrix}, \begin{pmatrix} 2 & 5 & 9 \\ 9 & 5 & 2 \end{pmatrix} \right\}.$$

2. The symmetric group S_{10} is the collection of all the permutations of the set $\{0,1,2,3,4,5,6,7,8,9\}$. An example of one of the many permutations in S_{10} is

$$f = \begin{pmatrix} 0 & 1 & 2 & 3 & 4 & 5 & 6 & 7 & 8 & 9 \\ 1 & 7 & 6 & 3 & 2 & 0 & 4 & 5 & 9 & 8 \end{pmatrix}$$

where $f(0) = 1$, $f(1) = 7$, $f(2) = 6$, and so on.

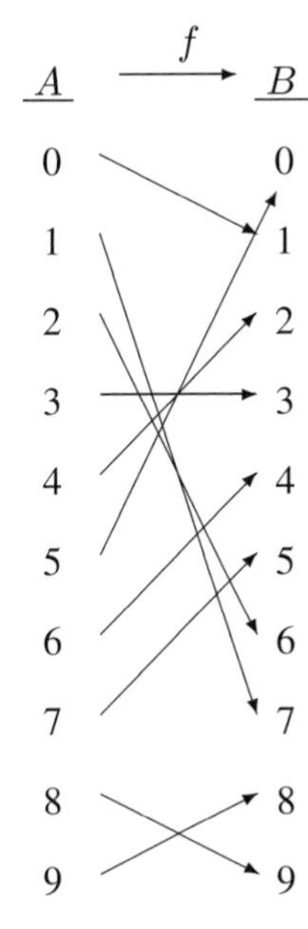

3. In Example 3.4.2, item (1) presents the permutation group S_1; item (2) presents the permutation group S_2; and item (3) presents the symmetric group S_3.

Cycles

Definition 3.4.5. *Let $a_1, a_2, a_3, \ldots, a_r$ be distinct members of the set $\{0, 1, 2, 3, \ldots, n-1\}$ where $n \geq 1$ and $1 \leq r \leq n$.*

1. *If $r = 1$, the* **cycle** *(a_1) is the permutation that sends $0 \to 0$, $1 \to 1, \ldots, n-1 \to n-1$ (i.e., it is the permutation that sends each number to itself).*
2. *If $r \geq 2$, the* **cycle** *$(a_1, a_2, a_3, \ldots, a_r)$ is the permutation from the symmetric group S_n that sends a_1 to a_2, a_2 to $a_3, \ldots, a_{r-1}$ to a_r, and a_r to a_1, and that sends each other element of the set $\{0, 1, 2, 3, \ldots, n-1\}$ to itself.*

$$a_1 \longrightarrow a_2 \longrightarrow a_3 \longrightarrow \quad \cdots \quad \longrightarrow a_{r-1} \longrightarrow a_r \quad (a_r \to a_1)$$

For example, in the symmetric group S_7 (the collection of permutations of the set $\{0, 1, 2, 3, 4, 5, 6\}$) the cycle $(0, 3, 5, 2, 4)$ is the permutation that sends $0 \to 3$, $3 \to 5$, $5 \to 2$, $2 \to 4$, $4 \to 0$, and all other elements to themselves ($1 \to 1$ and $6 \to 6$). This cycle can also be presented as

$$\begin{pmatrix} 0 & 1 & 2 & 3 & 4 & 5 & 6 \\ 3 & 1 & 4 & 5 & 0 & 2 & 6 \end{pmatrix}.$$

In the symmetric group S_4 (the collection of permutations of the set $\{0, 1, 2, 3\}$), the cycle $(0, 3, 1)$ is the permutation

$$\begin{pmatrix} 0 & 1 & 2 & 3 \\ 3 & 0 & 2 & 1 \end{pmatrix}.$$

If a cycle contains just one element, then it must be the cycle that sends *every* element to itself. The cycle (2) from the symmetric group S_4 is the permutation

$$\begin{pmatrix} 0 & 1 & 2 & 3 \\ 0 & 1 & 2 & 3 \end{pmatrix}.$$

The cycles (0), (1), and (3) from S_4 would also denote this permutation.

Now consider the permutation

$$f = \begin{pmatrix} 0 & 1 & 2 & 3 \\ 1 & 2 & 0 & 3 \end{pmatrix} \in S_3.$$

This permutation sends $0 \to 1$. It also sends $1 \to 2$. In a sense, applying f twice to the element 0 would send it to 2:

$$0 \xrightarrow{f} 1 \xrightarrow{f} 2.$$

Given that f also sends $2 \to 0$, applying f three times would send 0 back to 0:

$$0 \xrightarrow{f} 1 \xrightarrow{f} 2 \xrightarrow{f} 0.$$

This can be done for each element of the set $A = \{0, 1, 2, 3\}$ as illustrated next:

$$\begin{array}{ccccccc} \underline{A} & \xrightarrow{f} & \underline{A} & \xrightarrow{f} & \underline{A} & \xrightarrow{f} & \underline{A} \\ 0 & & 0 & & 0 & & 0 \\ 1 & & 1 & & 1 & & 1 \\ 2 & & 2 & & 2 & & 2 \\ 3 & \longrightarrow & 3 & \longrightarrow & 3 & \longrightarrow & 3 \end{array}$$

After applying f three times, each element from A gets sent back to itself. Thus,

$$0 \xrightarrow{f(0)=1 \quad f(1)=2 \quad f(2)=0} 0$$

$$1 \xrightarrow{f(1)=2 \quad f(2)=0 \quad f(0)=1} 1$$

$$2 \xrightarrow{f(2)=0 \quad f(0)=1 \quad f(1)=2} 2$$

$$3 \xrightarrow{f(3)=3 \quad f(3)=3 \quad f(3)=3} 3$$

Not only does each element eventually get sent back to itself, but by applying f three times, the set $A = \{0, 1, 2, 3\}$ has been divided into two smaller disjoint sets.

Consider the element $0 \in A$ and the *path* it takes. The path starts at 0 and then goes to 1 ($f(0) = 1$). The path then goes through 2 since $f(1) = 2$. There are no other elements in the path since $f(2) = 0$ and 0 is where the path started. Thus the elements $\{0, 1, 2\}$ are in the path that starts at 0. The same is true for the paths that start at 1 and 2. The only element from A not in $\{0, 1, 2\}$ is 3.

The path that starts at 3 is simple. Since $f(3) = 3$, the only element in it is 3. The element in the path starting at 3 is $\{3\}$. The two sets $\{0, 1, 2\}$ and $\{3\}$ have no elements in common (are *disjoint*), and are both subsets of A; their union equals A. In other words, they form a partition of A.

This behavior is not unique. Given any permutation in S_n, the paths it determines will divide $A = \{0, 1, 2, 3, \ldots, n-1\}$ into a partition of A. Consider the permutation

$$g = \begin{pmatrix} 0 & 1 & 2 & 3 & 4 & 5 & 6 & 7 \\ 3 & 4 & 0 & 5 & 1 & 2 & 7 & 6 \end{pmatrix} \in S_8.$$

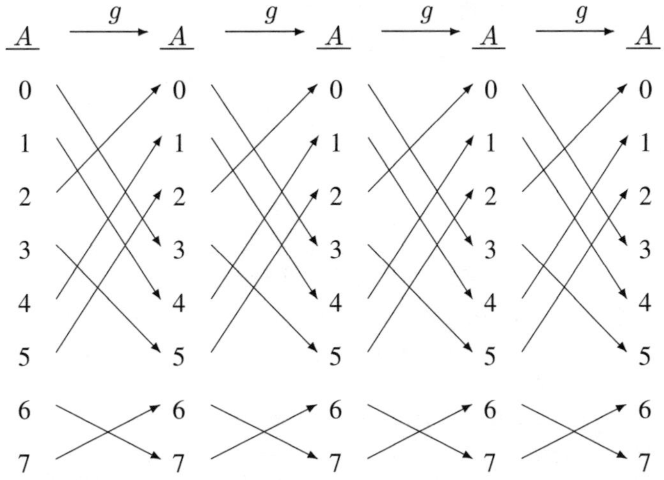

$$A \xrightarrow{g} A \xrightarrow{g} A \xrightarrow{g} A \xrightarrow{g} A$$

This permutation partitions the set $\{0, 1, 2, 3, 4, 5, 6, 7\}$ into the three subsets, $\{0, 2, 3, 5\}$, $\{1, 4\}$, and $\{6, 7\}$.

$$0 \xrightarrow{g(0)=3 \quad g(3)=5 \quad g(5)=2 \quad g(2)=0} 0$$

$$1 \xrightarrow{g(1)=4 \quad g(4)=1} 1$$

$$2 \xrightarrow{g(2)=0 \quad g(0)=3 \quad g(3)=5 \quad g(5)=2} 2$$

$$3 \xrightarrow{g(3)=5 \quad g(5)=2 \quad g(2)=0 \quad g(0)=3} 3$$

$$4 \xrightarrow{g(4)=1 \quad g(1)=4} 4$$

$$5 \xrightarrow{g(5)=2 \quad g(2)=0 \quad g(0)=3 \quad g(3)=5} 5$$

$$6 \xrightarrow{g(6)=7 \quad g(7)=6} 6$$

$$7 \xrightarrow{g(7)=6 \quad g(6)=7} 7$$

These two examples have demonstrated how a permutation from the symmetric group S_n will partition the set $\{0, 1, 2, 3, \ldots, n-1\}$ into a collection of disjoint subsets. Each subset from the partition was determined by a path created by the permutation.

In the first example,

$$f = \begin{pmatrix} 0 & 1 & 2 & 3 \\ 1 & 2 & 0 & 3 \end{pmatrix} \in S_3$$

yielded the partition $\{\{0, 1, 2\}, \{3\}\}$. It created $\{0, 1, 2\}$, since $f(0) = 1$, $f(1) = 2$, and $f(2) = 0$, and it created $\{3\}$, since $f(3) = 3$. Each of the paths that determine the partition can be represented by a cycle. Since $f(0) = 1$, $f(1) = 2$, and $f(2) = 0$, we can represent this by the cycle $(0, 1, 2)$. Since $f(3) = 3$, we can represent this by the cycle (3).

Order in the cycle is important. The first was listed as $(0, 1, 2)$ since $f(0) = 1$, $f(1) = 2$, and $f(2) = 0$. The second was listed as (3) since $f(3) = 3$. Combining the two cycles together gives the original permutation. This is written

$$f = \begin{pmatrix} 0 & 1 & 2 & 3 \\ 1 & 2 & 0 & 3 \end{pmatrix} = (0, 1, 2)(3).$$

In the second example,

$$g = \begin{pmatrix} 0 & 1 & 2 & 3 & 4 & 5 & 6 & 7 \\ 3 & 4 & 0 & 5 & 1 & 2 & 7 & 6 \end{pmatrix} \in S_8$$

yielded the partition $\{\{0, 2, 3, 5\}, \{1, 4\}, \{6, 7\}\}$. Given that $g(0) = 3$, $g(3) = 5$, $g(5) = 2$, and $g(2) = 0$, this can be represented by the cycle $(0, 3, 5, 2)$. The fact that $g(1) = 4$ and $g(4) = 1$ determines the cycle $(1, 4)$. And given that $g(6) = 7$ and $g(7) = 6$, this can be represented by the cycle $(6, 7)$. Again, order in the cycle is important. Combining these three cycles together gives the original permutation.

$$g = \begin{pmatrix} 0 & 1 & 2 & 3 & 4 & 5 & 6 & 7 \\ 3 & 4 & 0 & 5 & 1 & 2 & 7 & 6 \end{pmatrix} = (0, 3, 5, 2)(1, 4)(6, 7).$$

The combination of the cycles $(0, 3, 5, 2)$, $(1, 4)$, and $(6, 7)$ that results in permutation g is referred to as a *product of cycles*.

Theorem 3.4.6. *Every permutation from the symmetric group S_n is either a single cycle or the product of a finite number of disjoint cycles.*

Writing a permutation as a product of cycles is a more concise way to represent a permutation. All the information about the permutation is given in the cycles.

Example 3.4.7.

1. The two permutations below are from the symmetric group S_{10}, the set of permutations of the set $A = \{0, 1, 2, 3, 4, 5, 6, 7, 8, 9\}$.

   (a) $f = \left(\begin{smallmatrix} 0 & 1 & 2 & 3 & 4 & 5 & 6 & 7 & 8 & 9 \\ 9 & 2 & 8 & 6 & 0 & 5 & 1 & 7 & 3 & 4 \end{smallmatrix}\right) = (0, 9, 4)(1, 2, 8, 3, 6)(5)(7)$, since f partitions A into the subsets $\{0, 4, 9\}$, $\{1, 2, 3, 6, 8\}$, $\{5\}$, and $\{7\}$, and because $0 \xrightarrow{f} 9 \xrightarrow{f} 4 \xrightarrow{f} 0$, $1 \xrightarrow{f} 2 \xrightarrow{f} 8 \xrightarrow{f} 3 \xrightarrow{f} 6 \xrightarrow{f} 1$, $5 \xrightarrow{f} 5$, and $7 \xrightarrow{f} 7$.

(b) $g = \left(\begin{smallmatrix} 0 & 1 & 2 & 3 & 4 & 5 & 6 & 7 & 8 & 9 \\ 5 & 4 & 8 & 0 & 7 & 2 & 9 & 6 & 3 & 1 \end{smallmatrix}\right) = (0,5,2,8,3)(1,4,7,6,9)$, since g partitions A into the subsets $\{0,2,3,5,8\}$ and $\{1,4,6,7,9\}$, and because $0 \xrightarrow{g} 5 \xrightarrow{g} 2 \xrightarrow{g} 8 \xrightarrow{g} 3 \xrightarrow{g} 0$ and $1 \xrightarrow{g} 4 \xrightarrow{g} 7 \xrightarrow{g} 6 \xrightarrow{g} 9 \xrightarrow{g} 1$.

2. In the symmetric group S_5, the permutation

$$h = \begin{pmatrix} 0 & 1 & 2 & 3 & 4 \\ 3 & 4 & 1 & 2 & 0 \end{pmatrix} = (0,3,2,1,4)$$

is itself a cycle, as $0 \xrightarrow{h} 3 \xrightarrow{h} 2 \xrightarrow{h} 1 \xrightarrow{h} 4 \xrightarrow{h} 0$.

3. Given a product of cycles from the symmetric group S_n, the permutation can be obtained. In the symmetric group S_8,

$$(0,7,2,5)(1,4,3)(6) = \begin{pmatrix} 0 & 1 & 2 & 3 & 4 & 5 & 6 & 7 \\ 7 & 4 & 5 & 1 & 3 & 0 & 6 & 2 \end{pmatrix}.$$

Exercises 3.4

1. In this section, many examples of functions that are also permutations were given. Look in print or electronic media sources and find two examples of correspondences that result in permutations. Indicate what correspondence each function is establishing, why the function is one-to-one and onto (why the function is a permutation), and what the domain and range are. Also indicate, within the context or setting of the permutation, why it is important that the correspondence established results in a permutation. One way to address this is to indicate what might go wrong if the correspondence did not result in a permutation.

2. Determine which of the following functions from $A = \{1,2,3,4,5\}$ to $B = \{1,2,3,4,5\}$ are permutations. Make sure you indicate your answer and how you arrived at the answer you did.

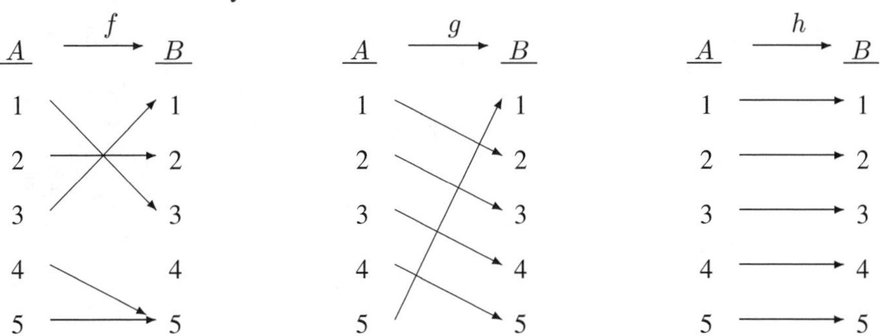

3. Consider the set $A = \{1,2,3\}$ and the set $B = \{7,8,9\}$. How many permutations from A to B are there? Be sure to document (write out) each one. How do you know that you have found them all and that there are not any more?

4. (a) Is there a permutation from the set $A = \{1,4,7,9\}$ into the set $B = \{2,3,4,9,10\}$? Explain how you got your answer.

(b) Is there a permutation from the set $A = \{0, 1, 2, 3, 4\}$ into the set $B = \{2, 4, 6, 8\}$? Explain how you got your answer.

(c) Given any finite set A, explain why there will always be at least one permutation $f : A \to A$. Be sure to indicate how that permutation would be obtained.

5. In this section, it was demonstrated that there is one permutation in the symmetric group S_1, two permutations in the symmetric group S_2, and six permutations in the symmetric group S_3. The goal of this exercise is to determine a formula to calculate how many permutations there are in a symmetric group S_n.

   (a) List all the permutations in the symmetric group S_4. How many are there?

   (b) For an integer $n \geq 1$, $n!$ (read *n factorial*) is the product $n\cdot(n-1)\cdot(n-2)\cdots 2\cdot 1$. For example, $5! = 5 \cdot 4 \cdot 3 \cdot 2 \cdot 1 = 120$. Compute $1!$, $2!$, $3!$, and $4!$.

   (c) Given your work in parts (a) and (b), find a formula that gives the number of permutations in the symmetric group S_n. Test it on S_1, S_2, S_3, and S_4.

   (d) Use your formula to determine the number of permutations in S_{10}.

6. Give two examples of elements from the symmetric group S_8.

7. Consider the following permutation.

$$f = \begin{pmatrix} 0 & 1 & 2 & 3 & 4 & 5 & 6 & 7 & 8 & 9 & 10 & 11 \\ 3 & 7 & 6 & 1 & 0 & 11 & 9 & 10 & 2 & 4 & 5 & 8 \end{pmatrix}$$

   (a) In which symmetric group is f a member?

   (b) Evaluate the expressions $f(1)$, $f(5)$, and $f(11)$.

8. Write each of these permutations from the symmetric group S_{10} in permutation notation.

A	$\xrightarrow{f}$	A	A	$\xrightarrow{g}$	A	h
0		0	0		0	$h(0) = 9$
1		1	1		1	$h(1) = 7$
2		2	2		2	$h(2) = 5$
3		3	3		3	$h(3) = 3$
4		4	4		4	$h(4) = 1$
5		5	5		5	$h(5) = 0$
6		6	6		6	$h(6) = 2$
7		7	7		7	$h(7) = 4$
8		8	8		8	$h(8) = 6$
9		9	9		9	$h(9) = 8$

9. Write each permutation as a product of cycles.

(a) $\begin{pmatrix} 0 & 1 & 2 & 3 & 4 & 5 & 6 & 7 & 8 & 9 \\ 3 & 4 & 5 & 6 & 7 & 8 & 9 & 1 & 2 & 0 \end{pmatrix}$ (b) $\begin{pmatrix} 0 & 1 & 2 & 3 & 4 & 5 & 6 & 7 & 8 & 9 \\ 2 & 1 & 4 & 5 & 6 & 3 & 8 & 9 & 0 & 7 \end{pmatrix}$

(c) $\begin{pmatrix} 0 & 1 & 2 & 3 & 4 & 5 & 6 \\ 2 & 3 & 4 & 5 & 6 & 0 & 1 \end{pmatrix}$ (d) $\begin{pmatrix} 0 & 1 & 2 & 3 & 4 \\ 2 & 3 & 4 & 1 & 0 \end{pmatrix}$

10. Consider the symmetric group S_7.

(a) If $f = \begin{pmatrix} 0 & 1 & 2 & 3 & 4 & 5 & 6 \\ 2 & 4 & 5 & 1 & 3 & 0 & 6 \end{pmatrix} \in S_7$, list the partition of $\{0, 1, 2, 3, 4, 5, 6\}$ that f determines.

(b) What is the largest number of subsets that an element of S_7 could partition $\{0, 1, 2, 3, 4, 5, 6\}$ into? What permutation in S_7 would it have to be? Write it as a product of cycles.

(c) What is the smallest number of subsets that an element of S_7 could partition $\{0, 1, 2, 3, 4, 5, 6\}$ into? Write down a permutation in S_7 that would give that partition. Write it as a single cycle.

Paper Assignments 3.4

1. **Defining.** This section included many examples of permutations, and earlier we studied a correspondence motivated by real data (the US national public debt) that was another example of a permutation. Look in print or electronic media sources and find two examples of correspondences that result in permutations. In a short paper, present each function, indicate what correspondence the function is establishing, why the function is one-to-one and onto, and what the domain and range are. Also indicate, within the context or setting of each permutation, why it is important that the correspondence established results in a permutation. One way to address this is to indicate what might go wrong if the correspondence did not result in a permutation. Use the work done on Exercise 3.4.1 as a base from which to begin this assignment.
2. **Argumentation.** Look in print or electronic media sources and find two correspondences, one that results in a permutation and one that results in a function that is not a permutation. In a short paper, present each function, explaining why one is a permutation and the other is not.
3. **Analysis.** Look in print or electronic media sources and find two correspondences, one that results in a permutation and one that results in a function that is not a permutation. In a short paper, present each function. Indicate why it is important for the one function to be a permutation and why it is not necessary for the other to be a permutation.
4. **Comparison.** Pick two sets of data from print or electronic sources, preferably from the same field (e.g., health, finance, sociology). Compare the sets in terms of how well the organization of the data enables you to derive the intended conclusions. In your comparison, you may want to apply the terms introduced in this section (permutations, cycles, partitions, etc.) and in previous chapters and sections.

Group Activities 3.4

1. Complete Exercise 3.4.5 as a group activity.
2. Consider the following permutations from the symmetric group S_{10}:

$$f = \begin{pmatrix} 0 & 1 & 2 & 3 & 4 & 5 & 6 & 7 & 8 & 9 \\ 2 & 4 & 5 & 1 & 9 & 8 & 6 & 7 & 0 & 3 \end{pmatrix} \quad g = \begin{pmatrix} 0 & 1 & 2 & 3 & 4 & 5 & 6 & 7 & 8 & 9 \\ 1 & 0 & 3 & 2 & 5 & 4 & 7 & 6 & 9 & 8 \end{pmatrix}$$

$$h = \begin{pmatrix} 0 & 1 & 2 & 3 & 4 & 5 & 6 & 7 & 8 & 9 \\ 1 & 2 & 3 & 4 & 5 & 6 & 7 & 8 & 0 & 9 \end{pmatrix} \quad k = \begin{pmatrix} 0 & 1 & 2 & 3 & 4 & 5 & 6 & 7 & 8 & 9 \\ 0 & 1 & 3 & 4 & 5 & 2 & 6 & 9 & 7 & 8 \end{pmatrix}$$

(a) Each member of the group chooses one of the permutations listed above and identifies the partition of $\{0, 1, 2, 3, 4, 5, 6, 7, 8, 9\}$ that it determines.

(b) As can be seen by the work done in part (a), different permutations determine different partitions of the set $\{0, 1, 2, 3, 4, 5, 6, 7, 8, 9\}$. With this in mind, what is the largest number of subsets that a permutation of S_{10} could partition the set $\{0, 1, 2, 3, 4, 5, 6, 7, 8, 9\}$ into? There is only one permutation of S_{10} that will do this. What is that permutation? Write it as a product of cycles.

(c) What is the smallest number of subsets that a permutation of S_{10} could partition the set $\{0, 1, 2, 3, 4, 5, 6, 7, 8, 9\}$ into? Many permutations in S_{10} can do this. Each member of the group should determine a distinct permutation in S_{10} that will give this partition, writing it down as a product of cycles. What are the similarities between all of these permutations?

Further Reading

Bernard, K. J., and Wellenzohn, H. J., *Foundations of Mathematics*, H&H Publishing Company, Clearwater, FL, 1997.

3.5 The IBM Scheme

Preliminary Activity. The UPC check digit scheme presented in Chapter 2 can be used only with identification numbers that consist of exactly 12 digits, and the ISBN scheme can be used only with identification numbers that consist of exactly ten digits. This is convenient in these cases because every Version A UPC consists of 12 digits and every ISBN has ten digits. Consider the following scenario:

> Your company wants to identify all of its products with its own identification number system. This system is to include a check digit scheme, but the identification numbers will be of varying lengths. For example, the system must handle numbers such as 1234 (length 4), 987654321 (length 9), and 1122334 (length 7) all at the same time. Develop a check digit scheme that your company can use.

Now that permutations have been firmly established, a check digit scheme that involves a permutation can be investigated. This more complicated check digit scheme, developed by

IBM, is used by credit card companies, libraries, blood banks, photo-finishing companies, pharmacies, some motor vehicle departments, and some German banks. It involves the permutation

$$\sigma = (0)(1,2,4,8,7,5)(3,6)(9)$$

from S_{10}. Using a permutation from S_{10} is a natural choice, since S_{10} is the collection of permutations of the set $\{0,1,2,3,4,5,6,7,8,9\}$, and each digit from an identification number is a digit from 0 to 9. It was for this very reason, the ability to apply a permutation to the digit 0, that we defined S_n to be the collection of permutations on the set $\{0,1,2,3,\ldots,n-1\}$. A nice aspect of the IBM scheme is that it can be used with identification numbers of any length.

Definition 3.5.1. The IBM Check Digit Scheme [3],[4]. *Let $a_1a_2a_3\ldots a_{n-1}$ represent an identifying number. The check digit a_n is appended to the number $a_1a_2a_3\ldots a_{n-1}$ to create the identification number $a_1a_2a_3\ldots a_{n-1}a_n$ by using the permutation*

$$\sigma = (0)(1,2,4,8,7,5)(3,6)(9)$$

in one of the following two ways.

1. *If n is even, the check digit a_n is assigned such that*

$$\sigma(a_1)+a_2+\sigma(a_3)+a_4+\cdots+\sigma(a_{n-1})+a_n = 0 \pmod{10}.$$

2. *If n is odd, the check digit a_n is assigned such that*

$$a_1+\sigma(a_2)+a_3+\sigma(a_4)+\cdots+\sigma(a_{n-1})+a_n = 0 \pmod{10}.$$

As mentioned above, many libraries use the IBM scheme to identify the books in their collections. The book *A Course in Number Theory and Cryptography* by Neal Koblitz [14] has been assigned the number 00001324136 9 by the Marist College library (see Figure 3.2). The check digit is the last digit, and it is separated from the identification number by a space.

The number associated with Koblitz's book has 12 digits in it, an even number. Consequently, to check whether it is a valid number, we use rule (1) in the definition of the IBM scheme, with $a_1a_2a_3a_4a_5a_6a_7a_8a_9a_{10}a_{11}a_{12} = 000013241369$ and $\sigma = (0)(1,2,4,8,7,5)(3,6)(9)$. This results in

$$\begin{aligned}
\sigma(a_1)+a_2+\sigma(a_3)+a_4+\sigma(a_5)+a_6+\sigma(a_7)+a_8+\sigma(a_9)+a_{10}&\\
+\sigma(a_{11})+a_{12} &= 0 \pmod{10}\\
\sigma(0)+0+\sigma(0)+0+\sigma(1)+3+\sigma(2)+4+\sigma(1)+3+\sigma(6)+9 &= 0 \pmod{10}\\
0+0+0+0+2+3+4+4+2+3+3+9 &= 0 \pmod{10}\\
30 &= 0 \pmod{10}.
\end{aligned}$$

FIGURE 3.2
Identification Number 00001324136 with Check Digit 9

Since $30 = 0 \pmod{10}$ is a true statement, this is a valid book number.

The Marist library book number $a_1a_2a_3a_4a_5a_6a_7a_8a_9a_{10}a_{11}a_{12} = 10011324136\ 9$ is not a valid number, as the following calculation shows:

$$\begin{aligned}\sigma(a_1)+a_2+\sigma(a_3)+a_4+\sigma(a_5)+a_6+\sigma(a_7)+a_8+\sigma(a_9)+a_{10}&\\ +\sigma(a_{11})+a_{12}&=0 \pmod{10}\\ \sigma(1)+0+\sigma(0)+1+\sigma(1)+3+\sigma(2)+4+\sigma(1)+3+\sigma(6)+9&=0 \pmod{10}\\ 2+0+0+1+2+3+4+4+2+3+3+9&=0 \pmod{10}\\ 33&=0 \pmod{10}.\end{aligned}$$

Since $33 = 0 \pmod{10}$ is a false statement ($33 = 3 \pmod{10}$), this is an invalid book number.

Before we can assign a check digit to an identification number, the identification number first needs to be generated. Suppose that a new book at the Marist library is to be assigned the 11-digit number $a_1a_2a_3a_4a_5a_6a_7a_8a_9a_{10}a_{11} = 21005620917$. The check digit C will be appended to the number to create the 12-digit book number 21005620917 C. To find the value for C, apply the IBM scheme to the entire number and determine what value of C will result in a true statement:

$$\begin{aligned}\sigma(a_1)+a_2+\sigma(a_3)+a_4+\sigma(a_5)+a_6+\sigma(a_7)+a_8+\sigma(a_9)+a_{10}&\\ +\sigma(a_{11})+a_{12}&=0 \pmod{10}\\ \sigma(2)+1+\sigma(0)+0+\sigma(5)+6+\sigma(2)+0+\sigma(9)+1+\sigma(7)+C&=0 \pmod{10}\\ 4+1+0+0+1+6+4+0+9+1+5+C&=0 \pmod{10}\\ 31+C&=0 \pmod{10}.\end{aligned}$$

The only digit C that will make this a true statement is $C = 9$, as $31 + C = 31 + 9 = 40 = 0 \pmod{10}$. Thus the book number is 21005620917 9.

The IBM scheme detects all single-digit errors. Listed below are two single-digit errors.

Correct Number:	21005**6**20917 9	21005620917 **9**
	↓	↓
Incorrect Number:	21005**3**20917 9	21005620917 **7**

Both errors are caught. The single-digit error involving the sixth digit is caught, as the following calculation results in a false statement:

$$\begin{aligned}\sigma(2)+1+\sigma(0)+0+\sigma(5)+3+\sigma(2)+0+\sigma(9)+1+\sigma(7)+9&=0 \pmod{10}\\ 4+1+0+0+1+3+4+0+9+1+5+9&=0 \pmod{10}\\ 37&=0 \pmod{10}.\end{aligned}$$

That equation is false, because $37 = 7 \pmod{10}$, not 0. The single digit error involving the check digit is also caught, as the following calculation results in the false statement $38 = 0 \pmod{10}$.

$$\begin{aligned}\sigma(2)+1+\sigma(0)+0+\sigma(5)+6+\sigma(2)+0+\sigma(9)+1+\sigma(7)+7&=0 \pmod{10}\\ 4+1+0+0+1+6+4+0+9+1+5+7&=0 \pmod{10}\\ 38&=0 \pmod{10}.\end{aligned}$$

It is very easy to show why this scheme catches all single-digit errors. Let $a_1 \ldots a_i \ldots a_n$ be an identification number with n even and $1 \le i \le n$. If n were odd, the same techniques could be used to show that all single-digit errors are caught. When a single-digit error occurs, $a_1 \ldots a_i \ldots a_n$ is transmitted as $a_1 \ldots b_i \ldots a_n$ with $a_i \neq b_i$ (a single-digit error where a_i is replaced by b_i).

Suppose this error were not caught. Then applying the IBM check digit scheme calculation to the correct and the incorrect number should result in two true statements. Now, there are two different situations that could occur when completing the IBM calculation. In the first case, $\sigma(a_i)$ and $\sigma(b_i)$ are needed to complete the calculation. In the second, simply a_i and b_i are needed.

Case I. In this case, both $\sigma(a_i)$ and $\sigma(b_i)$ are used in the calculation. Since both errors are not caught,

$$\sigma(a_1) + a_2 + \cdots + \sigma(a_i) + \cdots + \sigma(a_{n-1}) + a_n = 0 \pmod{10}$$

and

$$\sigma(a_1) + a_2 + \cdots + \sigma(b_i) + \cdots + \sigma(a_{n-1}) + a_n = 0 \pmod{10}.$$

This can also be written

$$\begin{aligned}&\big(\sigma(a_1) + a_2 + \cdots + \sigma(a_i) + \cdots + \sigma(a_{n-1}) + a_n\big)\\ &\qquad - \big(\sigma(a_1) + a_2 + \cdots + \sigma(b_i) + \cdots + \sigma(a_{n-1}) + a_n\big) = 0 \pmod{10}.\end{aligned}$$

This results in

$$\begin{aligned}0 &= \big(\sigma(a_1) + a_2 + \cdots + \sigma(a_i) + \cdots + \sigma(a_{n-1}) + a_n\big)\\ &\qquad - \big(\sigma(a_1) + a_2 + \cdots + \sigma(b_i) + \cdots + \sigma(a_{n-1}) + a_n\big) \pmod{10}\\ &= \sigma(a_1) + a_2 + \cdots + \sigma(a_i) + \cdots + \sigma(a_{n-1}) + a_n\\ &\qquad - \sigma(a_1) - a_2 - \cdots - \sigma(b_i) - \cdots - \sigma(a_{n-1}) - a_n \pmod{10}\\ &= \sigma(a_i) - \sigma(b_i) \pmod{10}.\end{aligned}$$

Thus $\sigma(a_i) - \sigma(b_i) = 0 \pmod{10}$. Since $a_i \neq b_i$ and σ is a permutation in S_{10}, $\sigma(a_i) - \sigma(b_i) = 0$. Adding $\sigma(b_i)$ to both sides yields the result $\sigma(a_i) = \sigma(b_i)$. Since σ is a permutation (in particular, one-to-one), this means that $a_i = b_i$. This is a contradiction. Thus the assumption that the error is not caught is false, and the error is caught.

Case II. In this case, simply a_i and b_i are used. Since both errors are not caught,

$$\sigma(a_1) + a_2 + \cdots + a_i + \cdots + \sigma(a_{n-1}) + a_n = 0 \pmod{10}$$

and

$$\sigma(a_1) + a_2 + \cdots + b_i + \cdots + \sigma(a_{n-1}) + a_n = 0 \pmod{10}.$$

This can also be written

$$\begin{aligned}&\big(\sigma(a_1) + a_2 + \cdots + a_i + \cdots + \sigma(a_{n-1}) + a_n\big)\\ &\qquad - \big(\sigma(a_1) + a_2 + \cdots + b_i + \cdots + \sigma(a_{n-1}) + a_n\big) = 0 \pmod{10}.\end{aligned}$$

This results in

$$\begin{aligned}
0 &= \big(\sigma(a_1) + a_2 + \cdots + a_i + \cdots + \sigma(a_{n-1}) + a_n\big) \\
&\qquad - \big(\sigma(a_1) + a_2 + \cdots + b_i + \cdots + \sigma(a_{n-1}) + a_n\big) \pmod{10} \\
&= \sigma(a_1) + a_2 + \cdots + a_i + \cdots + \sigma(a_{n-1}) + a_n \\
&\qquad - \sigma(a_1) - a_2 - \cdots - b_i - \cdots - \sigma(a_{n-1}) - a_n \pmod{10} \\
&= a_i - b_i \pmod{10}.
\end{aligned}$$

Thus $a_i - b_i = 0 \pmod{10}$. Since $a_i \neq b_i$ and both a_i and b_i are integers between 0 and 10, $a_i - b_i = 0$. Adding b_i to both sides, the calculation results in $a_i = b_i$. This is a contradiction. Thus the assumption that the error is not caught is false and the error is caught.

Although the IBM scheme does catch all single-digit errors, it does not catch all transposition-of-adjacent-digits errors. It catches all transposition errors except those where the adjacent digits 0 and 9 are transposed.[1] Recall the valid Marist College library book number 21005620917 9. Listed below are two transposition-of-adjacent-digits errors. The first one will be caught, but the second will not, as the second error involves 0 and 9.

Correct Number:	21005**62**0917 9	210056**209**17 9
	↓	↓
Incorrect Number:	21005**26**0917 9	2100562**90**17 9

The first error is caught, as the following calculation indicates:

$$\begin{aligned}
\sigma(2) + 1 + \sigma(0) + 0 + \sigma(5) + 2 + \sigma(6) + 0 + \sigma(9) + 1 \\
+\sigma(7) + 9 &= 0 \pmod{10} \\
4 + 1 + 0 + 0 + 1 + 2 + 3 + 0 + 9 + 1 + 5 + 9 &= 0 \pmod{10} \\
35 &= 0 \pmod{10} \quad \textit{False.}
\end{aligned}$$

The second error is not caught, as the following calculation indicates:

$$\begin{aligned}
\sigma(2) + 1 + \sigma(0) + 0 + \sigma(5) + 6 + \sigma(2) + 9 + \sigma(0) + 1 \\
+\sigma(7) + 9 &= 0 \pmod{10} \\
4 + 1 + 0 + 0 + 1 + 6 + 4 + 9 + 0 + 1 + 5 + 9 &= 0 \pmod{10} \\
40 &= 0 \pmod{10} \quad \textit{True.}
\end{aligned}$$

Exercises 3.5

1. Suppose that the following two numbers are Marist College library book identification numbers that have just been scanned into the library computer. Using the IBM check digit scheme, determine which number is valid and which is invalid. Remember that the last digit is the check digit.

   00001301769 8 13092278172 4

[1] The reason for this is found in Group Activity 3.5.1.

2. Suppose that the following two numbers are library book identification numbers that have just been scanned into a library's computer. Using the IBM check digit scheme, determine which number is valid and which is invalid. Remember that the last digit is the check digit.

   025003104756 7 230090768142 2

3. The number 00100673283 will be used to identify a book. Using the IBM check digit scheme, assign a check digit to this number. Be sure to identify the check digit and then to write out the entire identification number (including the check digit).

4. The IBM check digit scheme is also used by credit card companies. Listed below are two credit card numbers. Use the IBM scheme to determine which number is valid and which is invalid. The last digit in each number is the check digit.

   1031 2286 9302 4422 9234 1001 2396 3178

5. The IBM check digit scheme will catch all transposition-of-adjacent-digits errors, except those that involve 0 and 9. Using the IBM scheme, all transposition errors will be caught for what types of identification numbers? What type of identification numbers will have no transposition-of-adjacent-digits errors caught?

Paper Assignment 3.5

1. **Comparison.** You are employed by an organization that wants to use a check digit scheme. By now, you are familiar with a variety of systems. An initial survey done by the company has narrowed down the choices to the ISBN and the IBM schemes. Your task is to recommend one of them to the company by writing a report, called a feasibility study. Compare the two schemes, paying attention to the following factors:

   (a) ease of use (the length of time involved with and complexity of the calculations);

   (b) strength (in terms of the errors it can detect);

   (c) flexibility (the ISBN scheme is designed for an identification number of a fixed length while the IBM scheme can be manipulated and adjusted to numbers of any length).

   At the end of this comparison, recommend one of the schemes and state your reasons for doing so.

Group Activity 3.5

1. The IBM scheme does not catch all transposition-of-adjacent-digits errors. If a and b are two adjacent digits in a UPC, the transposition error $\dots ab \dots \to \dots ba \dots$ is not caught when $a = 0$ and $b = 9$ or when $a = 9$ and $b = 0$. The goal of this activity

is to determine why this happens. Consider the following transposition-of-adjacent-digits errors involving four valid identification numbers. Each member of each group chooses one valid number and investigates the two associated errors.

Correct Number	Incorrect Number A	Incorrect Number B
19032 2	91032 2	10932 2
09186 8	09168 8	90186 8
21149 0	21419 0	21140 9
09901 0	09910 0	09091 0

(a) Making sure to write out all the calculations, show that the Correct Number is actually a valid identification number.

(b) Making sure to write out all the calculations, show that the IBM scheme will catch the transposition-of-adjacent-digits error in Incorrect Number A but not that in Incorrect Number B.

(c) In parts (a) and (b), the following was computed for each number, where $a_1a_2a_3a_4a_5a_6$ was the Correct or Incorrect Number:

$$\sigma(a_1) + a_2 + \sigma(a_3) + a_4 + \sigma(a_5) + a_6 = 0 \pmod{10}.$$

Compare the calculation

$$\sigma(a_1) + a_2 + \sigma(a_3) + a_4 + \sigma(a_5) + a_6$$

done for the Correct Number with Incorrect Number A and with Incorrect Number B. In both cases, there should be only a minor difference between the pair of calculations. What is it in each case? Might the permutation σ shed some light on your investigation?

(d) Given your work in the previous steps, conjecture as to why the transposition of adjacent digits a and b ($\ldots ab \ldots \rightarrow \ldots ba \ldots$) is not caught when $a = 0$ and $b = 9$ or when $a = 9$ and $b = 0$.

Further Reading

Gallian, J. A., The Mathematics of Identification Numbers, *College Mathematics Journal*, 22(3), 1991, 194–202.

Gallian, J. A., Error Detection Methods, *ACM Computing Surveys*, 28(3), 1996, 504–517.

3.6 Graphs of Functions

Preliminary Activity. Consider the data presented in Table 3.7 on the number of homicides in the city of New York for the years 1985 through 1996. Study the data and find a way to present them visually so that the trends in the homicide rate are easily seen. Based on this visual presentation, explain, as best you can, the trends found in the data.

TABLE 3.7
Homicides in New York City

Year	Homicides	Year	Homicides
1985	1384	1991	2154
1986	1582	1992	1995
1987	1672	1993	1946
1988	1892	1994	1561
1989	1905	1995	1177
1990	2245	1996	983

Table 3.8 lists the daily volume (total number of stocks traded, in millions) on the New York Stock Exchange (NYSE) between March 21, 1998, and June 9, 1998. This correspondence between date and volume forms a function. Use Table 3.8 to answer the following series of questions.

TABLE 3.8
NYSE Volume (in Millions) from March 21–June 9, 1998

Date	Volume	Date	Volume	Date	Volume	Date	Volume
March 21	0	April 11	0	May 1	581.3	May 21	551.8
March 22	0	April 12	0	May 2	0	May 22	444.5
March 23	630.8	April 13	565.9	May 3	0	May 23	0
March 24	614.0	April 14	613.2	May 4	551.3	May 24	0
March 25	676.2	April 15	684.1	May 5	583.4	May 25	0
March 26	606.4	April 16	698.5	May 6	608.3	May 26	541.1
March 27	581.9	April 17	671.6	May 7	591.8	May 27	703.3
March 28	0	April 18	0	May 8	567.6	May 28	588.7
March 29	0	April 19	0	May 9	0	May 29	556.5
March 30	497.2	April 20	596.7	May 10	0	May 30	0
March 31	674.6	April 21	674.6	May 11	560.6	May 31	0
April 1	677.0	April 22	694.8	May 12	604.2	June 1	542.4
April 2	673.7	April 23	649.9	May 13	602.2	June 2	594.6
April 3	653.2	April 24	630.3	May 14	578.1	June 3	584.1
April 4	0	April 25	0	May 15	621.7	June 4	577.0
April 5	0	April 26	0	May 16	0	June 5	558.1
April 6	628.6	April 27	691.6	May 17	0	June 6	0
April 7	670.4	April 28	691.0	May 18	519.8	June 7	0
April 8	616.0	April 29	653.3	May 19	570.2	June 8	543.0
April 9	548.7	April 30	704.4	May 20	597.1	June 9	563.1
April 10	0						

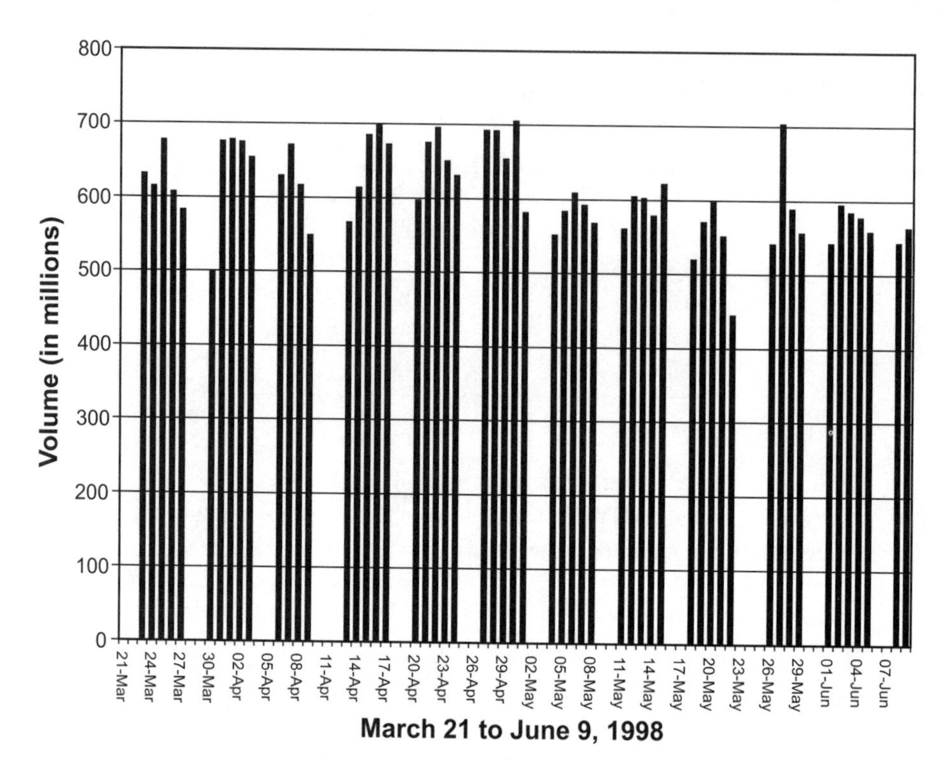

FIGURE 3.3
NYSE Volume (in Millions) from March 21–June 9, 1998

1. On what day(s) was the largest number of shares traded?
2. On what day(s) was the smallest number of shares traded?
3. Ignoring holidays and weekends (when trading = 0), when was the most dramatic increase and decrease in the volume of shares traded?
4. Between March 21, 1998, and June 9, 1998, was there a general increase or decrease in volume, or did it remain about the same?

How long did it take you to answer all four questions? Five minutes, ten minutes, or more? It may have taken you a long time to sort through and compare all of the numbers. These data are much easier to read when presented in graphical form, as Figure 3.3 illustrates. In this form, all four questions can be answered in about one minute.

Information of the kind asked for above can be obtained much more quickly when the data are presented in graphical form. A graph visually illustrates trends not seen when the data are presented in aggregate form. However, to perform a more detailed analysis or to obtain precise values, the original data, such as those in Table 3.8, should be used.

Any data that present a correspondence that forms a function (*functional data*) can be displayed in some sort of graphical form. The different possibilities cannot all be listed here, but there are some general guidelines.

When graphing a function $f : A \to B$ for functional data, two axes are used. Each axis is marked in increments where the distance between each mark represents one unit.

Sometimes an arrow is put at one end to indicate the direction in which the number of units is getting larger. Each mark on the axis measures a certain number of units. A unit could be a day (as in Figure 3.3), a numerical quantity $(1, 10, 100$, etc.), a month, a year, and so on.

$\ldots \quad 0 \quad 1 \quad 2 \quad 3 \quad 4 \quad \ldots$

One axis, usually the horizontal one or the *x-axis*, is used to list the elements from the domain, and the other axis, usually the vertical one or the *y-axis*, is used to list elements from the range. The numbers or elements listed along the x-axis are called *x-coordinates*, and the elements listed along the y-axis are called *y-coordinates*. In the graph of the NYSE volume, the x-coordinates are the days between March 21 and June 9, 1998. The days are listed in chronological order. Each unit on the vertical axis, or each y-coordinate, corresponds to 100 million. The marks go from 0 to 800. The graph of the volume of the NYSE, presented in Figure 3.3 is an example of a *bar graph* or *bar chart*.

There are many different types of graphs. Another type, shown in Figure 3.4, presents the US national public debt, in billions of dollars, for the years 1978 through 1997. The precise data can be found in Table 3.4.

Figure 3.4 is an example of a *timeplot*. In a timeplot, the domain is always time related. It could be a list of hours, days, weeks, or years. The range values are plotted against the time at which they were observed. This is done by drawing a dot on the graph. Once all the values are plotted, a line is drawn between each adjacent pair of points to help show the patterns in the data.

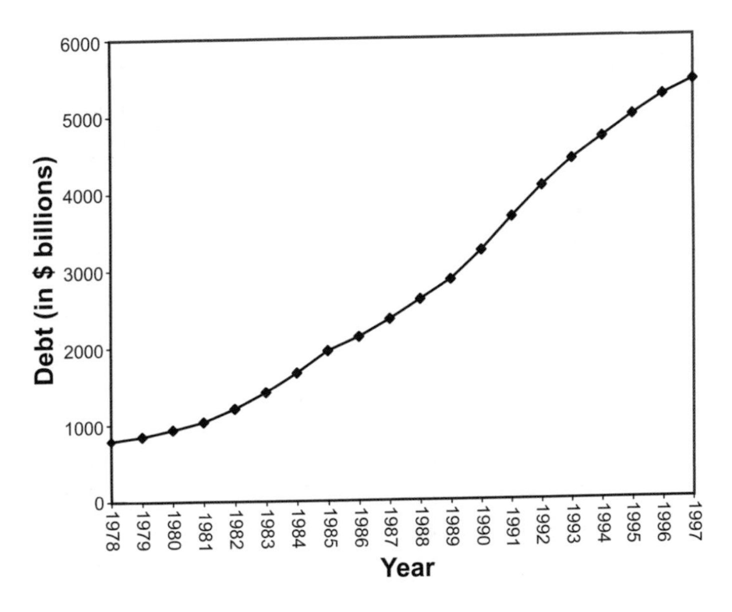

FIGURE 3.4
Timeplot of the US National Public Debt

For the US national public debt, the function f that is being represented corresponds each year to the size, in billions of dollars, of the US national public debt that year. In other words, dom(f) is the years from 1978 to 1997 or $\{1978, 1979, 1980, \ldots, 1997\}$ and the ran(f) is the set of numerical values between 0 and 6000 (billions of dollars). Each unit on the horizontal axis corresponds to one year. Each unit on the vertical axis corresponds to 1000, and each number indicates billions of dollars.

The graphs in Figures 3.3 and 3.4 present data in a visual way. This visualization makes it easier to spot trends in the data. For example, from the graph in Figure 3.3, it is readily seen that the lowest volume day for the NYSE was May 22 and the highest volume day was April 22. While the volume traded fluctuated from day to day, the overall amount of shares traded remained fairly consistent from week to week. This might indicate that stock prices were fairly stable during this period, as no major sell-offs appeared.

From the graph presented in Figure 3.4, it can easily be determined that the US national public debt is growing at a very fast rate. In fact, when presented this way, it appears as if the national debt will grow without end. This alarming picture might motivate our government to curtail its deficit spending practices.

Exercises 3.6

1. Look at print or electronic media sources to find three examples of graphs presenting functional data. For each graph indicate the correspondence that it is presenting, identify the two sets of data being used in the correspondence, and show how individual pieces of data can be obtained from the chart or graph.
2. The data presented in Table 3.9 gives the composite index of the New York Stock Exchange (NYSE) for the month of September 1999 (only for the days the NYSE was open). The composite index gives an across-the-board measure of the value of the market, as compared to a measure of 50.00 established in 1966.

TABLE 3.9
NYSE Composite Index for September 1–30, 1999

Day of Month	Composite Index	Day of Month	Composite Index	Day of Month	Composite Index
1	617.29	13	620.81	22	601.16
2	611.33	14	615.11	23	590.98
3	625.89	15	608.47	24	589.54
7	623.71	16	607.32	27	591.62
8	621.87	17	613.10	28	590.08
9	621.99	20	612.10	29	586.24
10	622.90	21	600.99	30	592.79

(a) Present these data using both a bar chart and a timeplot. Start numbering the vertical y-axis at 550.

(b) Using the graphs you have just created, analyze the behavior of the NYSE during September 1999. Do you notice any trends? What were the highs and lows? Why might they have occurred then? What might have occurred on these dates?

Paper Assignments 3.6

1. **Analysis.** When presented in aggregate form, data are often hard to read, understand, and interpret. Presenting data sets graphically illuminates them by showing trends and yielding information not apparent otherwise. In this assignment you will present data in a form appropriate for analysis and then discuss your findings in writing.

   First pick an area of interest. Search available databases (e.g., the Web, the library, online databases, CD-ROM, microfilm) and collect data on the topic you have chosen to investigate. Arrange and present the raw data in an organized, but aggregate, format. Consider the data from a mathematical point of view: What sets are involved? Is there a correspondence that is being established? If so, does this correspondence form a function? What are the domain and range of this function? Is it one-to-one or onto? Does the correspondence form a permutation?

   In your paper, present the data in graphical form in a way that helps the reader arrive at the same conclusions as you. Provide captions for each illustration and include a brief description (e.g., Graph 1. Skateboard Sales). Remember to describe in words what the illustrations show. What trends and information can the reader see when the data are presented in this way? Indicate your objective clearly. For instance, are you trying to identify a problem, solve a problem, explore history, or predict a trend?

Group Activity 3.6

1. The government, corporations, researchers, and numerous other organizations and individuals use graphs as a way to present data. However, graphs can be manipulated in order to downplay or exaggerate trends in the data.

   For example, consider some data on the Ford Motor Company. The price for a share of Ford stock fell from \$64.50 to \$56.00 between May 12 and May 24, 1999. The raw data for the approximated closing stock price for the working days from May 12 through May 24 are listed in Table 3.10. Using the data presented in Table 3.10, construct four timeplots as directed below.

   (a) Let the x-axis (horizontal axis) list the dates given and the y-axis (vertical axis) list the stock prices in units of \$10. In other words, start the y-coordinates at 0 and go up by 10 each time.

   (b) Let the x-axis list the dates given and the y-axis list the stock prices in units of \$5. In other words, start the y-coordinates at 0 and go up by 5 each time.

TABLE 3.10
Ford Stock Closing Value, May 12–24, 1999

Day of Month	Stock Price (Dollars)	Day of Month	Stock Price (Dollars)
12	64.50	19	57.60
13	63.00	20	57.90
14	60.90	21	58.10
17	57.49	24	56.00
18	58.20		

(c) Let the x-axis list the dates given and the y-axis list the stock prices in units of $1. Start the hash marks at 50 and go up by 1 each time. (Starting at 0 would not leave enough room on your paper to draw the timeplot.)

(d) Let the x-axis list the dates given and the y-axis list the stock prices in units of $.50 (50 cents). Start the hash marks at 55 and go up by 0.50 each time. (Starting at 0 would not leave enough room on your paper to draw the timeplot.)

As a group, compare the four graphs. Which portrays the decrease in stock prices in the most dramatic way? Which portrays the decrease in the least dramatic way? Which graph might Ford want to show to shareholders to convince them not to sell their stock? Which graph would Ford executives want to show to the Board of Directors to convince them that some drastic cost-cutting initiatives are needed?

Further Reading

Richardson, D., and St. John, P., Plotting to Succeed, *Geographical Magazine*, 64(12), 1992, 42–44.

Harris, R., The Power of Information Graphics, *IIE Solutions*, 31, 1999, 26–27.

4

Symmetry and Rigid Motions

Of the check digit schemes presented so far, only the ISBN scheme catches all single-digit and transposition-of-adjacent-digits errors. However, to do this, the scheme had to introduce the character X, a non-digit, into the identification number system. Since then, the goal has been to develop a scheme that catches all single-digit and transposition-of-adjacent-digits errors without using a non-digit character. The work done in this chapter on symmetry and rigid motions will set the stage for the creation of such a check digit scheme. In addition, the notation established in Chapter 3 for permutations will be used in a mathematical investigation of symmetry.

4.1 Symmetry

Preliminary Activity. Recall or bring to class two familiar shapes that "exhibit symmetry" and two that do not. For each shape write a brief description as to why it does or does not exhibit symmetry.

When an object or picture is said to be *symmetric*, this usually implies that it is uniformly shaped or has a balanced arrangement. The word "symmetry" comes from an ancient Greek word meaning "the same measure." The Greeks believed that exhibiting symmetry or proportion was a virtue to strive for. Symmetry is everywhere: in architecture, nature, art, sports, and many other places. Symmetry also plays a central role in mathematics.

Consider the six shapes below. Are they symmetric?

If an item is symmetric, what are the characteristics that make it so? If an item is not symmetric, what are the characteristics that prevent it from being so?

 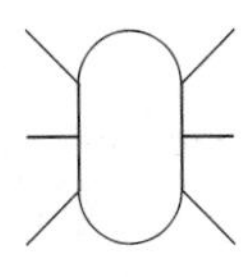 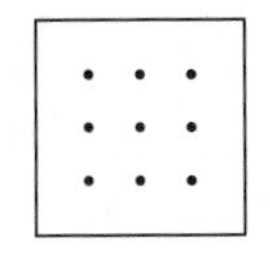 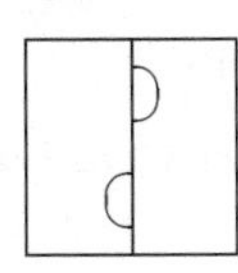 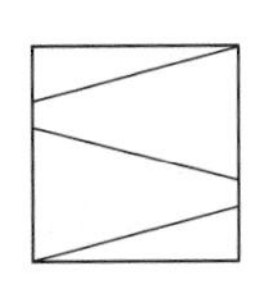 

What is meant when an object or picture is said to be symmetric? This notion needs to be more specifically defined. The most basic type of symmetry is called *reflective* symmetry. An object displays this type of symmetry if a line can be drawn through the object, creating two halves that are identical in size and design. Put another way, when an object exhibits reflective symmetry, one half of the object is a mirror image of the other half of the object. Each image in Figure 4.1 exhibits reflective symmetry. For each, the *line of symmetry* is a vertical line drawn down the middle of the image.

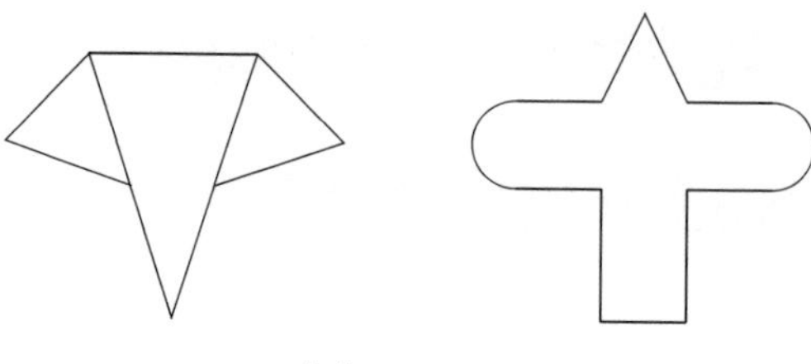

FIGURE 4.1
Examples of Reflective Symmetry

In addition to reflective symmetry, there is also *rotational* symmetry. An object exhibits this type of symmetry if, when it is rotated a certain number of degrees around a focal point, the object appears unchanged. The images in Figure 4.2 each exhibit rotational symmetry. For the shape on the left, the point of rotation is in the center of the star. For the shape on the right, the point of rotation is the center of the image where the two lines cross.

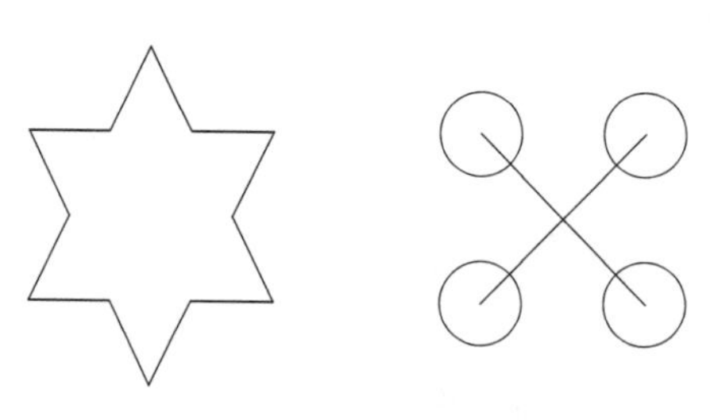

FIGURE 4.2
Examples of Rotational Symmetry

The mathematical notion of symmetry continues in the same vein.

Definition 4.1.1. *Given an object, a* ***symmetry*** *of the object is a way to reflect its image over a line, rotate its image over a point, or move its image any other way so that this movement does not change the size, position (orientation) in space, or visual presentation of the object.*

In other words, after a symmetry has been applied to an object there will be no change in the visual presentation of the object. Consider the rectangle in Figure 4.3 with each corner or *vertex* labeled a, b, c, and d. A reflective symmetry is applied to it by reflecting the rectangle over line l. While the object has moved, as indicated by the new positions of the vertices a, b, c, and d (since the rectangle has been reflected over line l, a and b, along with c and d, have changed places), the visual image of the rectangle has not changed.

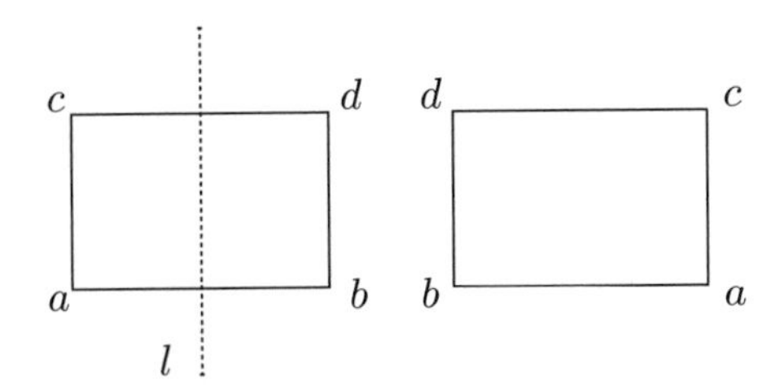

FIGURE 4.3
An Example of Reflective Symmetry on a Rectangle

The permutation notation developed in Chapter 3 is used here to denote the reflective symmetry of the rectangle shown in Figure 4.3. The rectangle on the left will be called the *original* position of the rectangle. While the reflective symmetry, by definition, did not change the visual presentation of the image, it did permute the rectangle's corners or vertices. Thus corner a moved to where corner b originally was or $a \to b$. Corner b moved to where corner a originally was or $b \to a$. Corner c moved to where corner d originally was or $c \to d$. Corner d moved to where corner c originally was or $d \to c$. This can be represented concisely as an array, where the top row lists the original positions and the bottom row lists the positions to which each corresponding corner moved:

$$\begin{pmatrix} a & b & c & d \\ b & a & d & c \end{pmatrix}.$$

The simplest shapes to study with regard to symmetries are *regular n-gons*. The following definitions will set the stage for investigating the symmetries of these shapes.

Definition 4.1.2. *A* ***polygon*** *is formed by a collection of line segments, called* **sides**, *such that each side intersects exactly two other sides, one at each endpoint and in no other places.* (See Figure 4.4.)

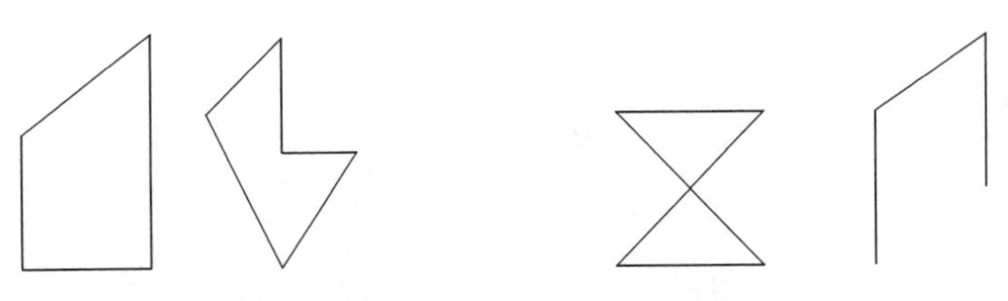

FIGURE 4.4
Left, Polygons. *Right,* Not Polygons

Definition 4.1.3. *A polygon is **convex** when the line drawn between any two points in the interior of the polygon does not contain any points from the exterior of the polygon.* (See Figure 4.5.)

FIGURE 4.5
Left, Convex Polygons. *Right,* Non-Convex Polygons

Definition 4.1.4. *A **regular polygon** is a convex polygon that has all sides of equal length and all angles of equal measure. If a regular polygon has n sides, with $n \geq 3$, it is referred to as a **regular n-gon**.*

To begin the mathematical study of symmetries, all the symmetries of a regular 5-gon (a pentagon) will be found. A pentagon exhibits both reflective and rotational symmetries. Note that the first symmetry in Figure 4.6, denoted id (for Latin *idem*, "the same"), is the trivial one, where the object is not moved at all. The abbreviation rf denotes a reflective symmetry, and rt denotes a rotational symmetry.

Examine the ten symmetries of a pentagon illustrated in Figure 4.6. There are the trivial symmetry id, four rotational symmetries (rt_1, rt_2, rt_3, and rt_4), and five reflective symmetries (rf_1, rf_2, rf_3, rf_4, and rf_5).

Each symmetry can be represented, using permutation notation, as a permutation of the vertices a, b, c, d, and e:

$$id = \begin{pmatrix} a & b & c & d & e \\ a & b & c & d & e \end{pmatrix}, \ rt_1 = \begin{pmatrix} a & b & c & d & e \\ b & c & d & e & a \end{pmatrix}, \ rt_2 = \begin{pmatrix} a & b & c & d & e \\ c & d & e & a & b \end{pmatrix},$$

$$rt_3 = \begin{pmatrix} a & b & c & d & e \\ d & e & a & b & c \end{pmatrix}, \ rt_4 = \begin{pmatrix} a & b & c & d & e \\ e & a & b & c & d \end{pmatrix}, rf_1 = \begin{pmatrix} a & b & c & d & e \\ b & a & e & d & c \end{pmatrix},$$

$$rf_2 = \begin{pmatrix} a & b & c & d & e \\ d & c & b & a & e \end{pmatrix}, rf_3 = \begin{pmatrix} a & b & c & d & e \\ a & e & d & c & b \end{pmatrix}, rf_4 = \begin{pmatrix} a & b & c & d & e \\ c & b & a & e & d \end{pmatrix},$$

$$rf_5 = \begin{pmatrix} a & b & c & d & e \\ e & d & c & b & a \end{pmatrix}.$$

This collection of symmetries of a regular 5-gon is called the *dihedral group* D_{10}.

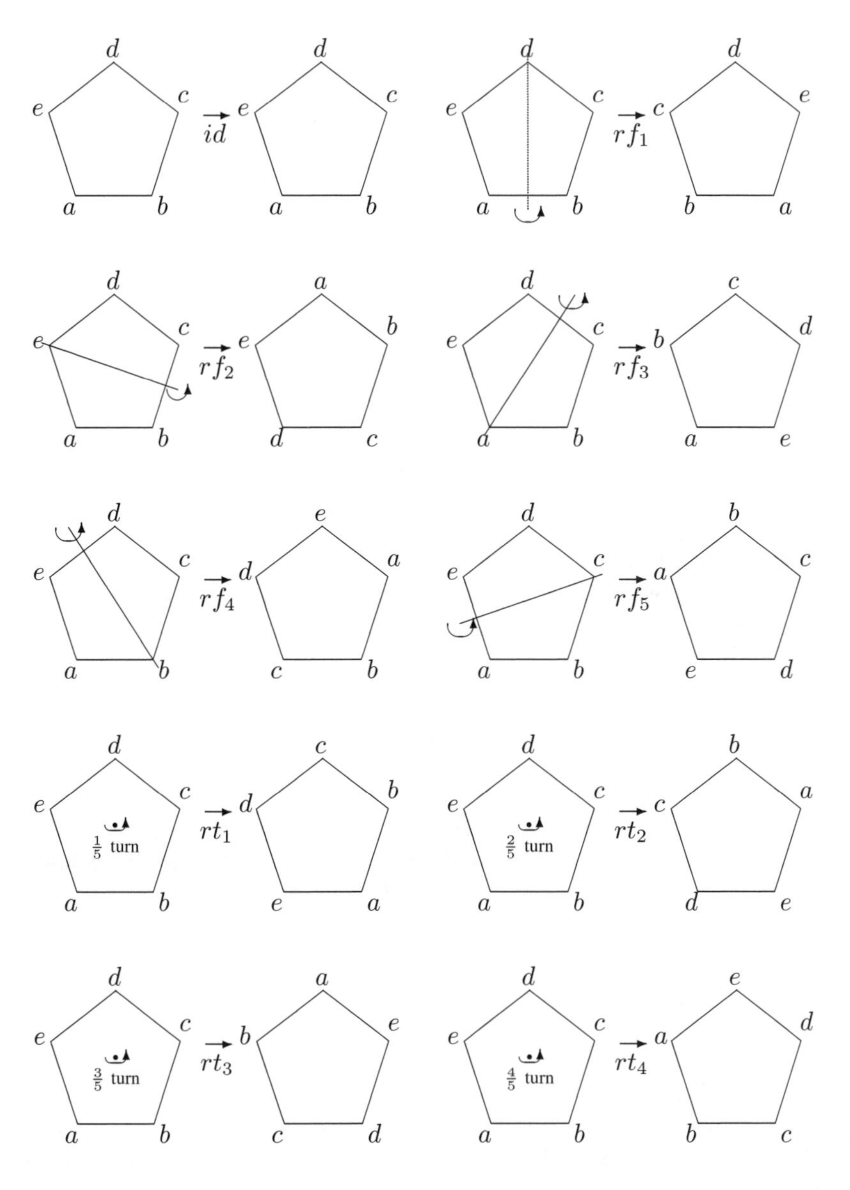

FIGURE 4.6
Symmetries of a Regular 5-gon

The search for the symmetries of geometric shapes extends far beyond the regular 5-gon (the pentagon). In mathematics, the collections of symmetries of a variety of different geometric shapes are studied. The easiest ones to determine are the collections of symmetries of regular n-gons.

Definition 4.1.5. *The collection of symmetries of a regular n-gon, for $n \geq 3$, is called the **dihedral group** D_{2n}.*

For example, the dihedral group D_6 is the collection of symmetries of a regular 3-gon (an equilateral triangle), and the dihedral group D_8 is the collection of symmetries of a regular 4-gon (a square).

Exercises 4.1

1. List all the reflective and rotational symmetries of the following pictures. Be sure to identify each symmetry and how the image exhibits that symmetry.

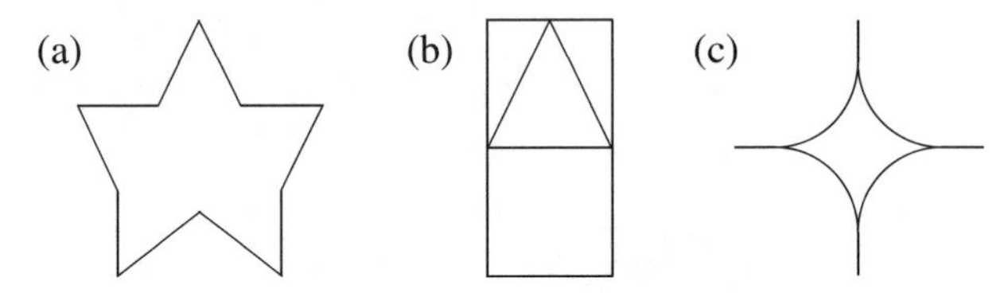

2. In your own words, define a polygon. Draw some examples of geometric shapes that are and are not polygons. Find at least three real-world examples of polygons, and find at least three real-world examples of non-polygons.
3. In your own words, define a convex polygon. Draw some examples of geometric shapes that are and are not convex polygons. Find at least three real-world examples of convex polygons, and find at least three real-world examples of polygons that are not convex.
4. Find all the symmetries of the following geometric shapes. Be sure to list all the symmetries for each object and to show how the object exhibits those symmetries.

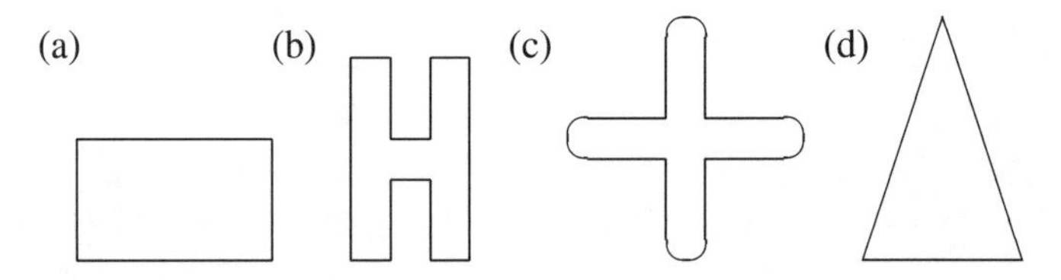

5. Go to the library and look at a few books of art (paintings, sculpture, architecture, etc.). Photocopy five different items and find all the symmetries in each one. Be sure to present each item and to describe all of the symmetries each item exhibits.
6. Go to the parking lot and find three different hubcap designs. Copy each design onto a sheet of paper and then find all of its symmetries. Be sure to identify each symmetry and how the object exhibits that symmetry.
7. (a) List all the members of the dihedral group D_6, the collection of symmetries on a regular 3-gon (an equilateral triangle). Be sure to identify each symmetry and how the 3-gon exhibits each of the symmetries you listed.

   (b) List all the members of the dihedral group D_8, the collection of symmetries on a regular 4-gon (a square). Be sure to identify each symmetry and how the 4-gon exhibits each of the symmetries you listed.

(c) List all the members of the dihedral group D_{12}, the collection of symmetries on a regular 6-gon or (a hexagon). Be sure to identify each symmetry and how the 6-gon exhibits each of the symmetries you listed.

8. Find two objects, images, or shapes that have the same number and type of symmetries as the dihedral group D_6. Explain why the symmetries for the shapes you found are the same as those in D_6.

9. Find two objects, images, or shapes that have the same number and type of symmetries as the dihedral group D_8. Explain why the symmetries for the shapes you found are the same as those in D_8.

10. Can an object exist that has a nontrivial rotational symmetry, but no reflective symmetry? If so, draw the image. If not, give a brief description of why this is not possible.

Paper Assignment 4.1

1. **Analysis.** Symmetry exists in nature, art, and architecture. Find an example from one of these areas that exhibits reflective and rotational symmetry. In a short essay present the item you have chosen and describe all of the reflective and rotational symmetries that appear in it. Discuss the function(s), if any, of these symmetries (e.g., to emphasize harmony, to define beauty in art). Include any diagrams that will help illustrate the points that you are making in your paper.

Group Activity 4.1

1. In this section, all of the symmetries of a pentagon (regular 5-gon) were discovered. This was part of a general discussion of symmetries for planar or two-dimensional objects. However, symmetries of three-dimensional objects are also interesting. Consider the two three-dimensional objects (a cube and a tetrahedron) presented below. By using reflections and rotations, all of the symmetries of each figure can be obtained.

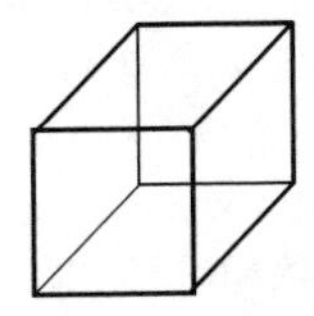
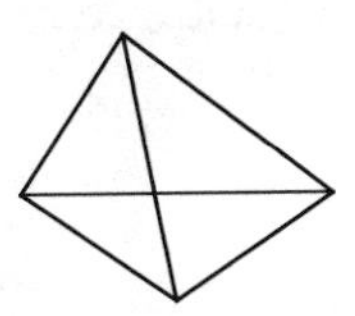

(a) Label each of the eight vertices of the cube with the lowercase letters *a, b, c, d, e, f, g,* and *h*, and label the four vertices of the tetrahedron with the lowercase letters *a, b, c,* and *d*.

(b) Find all of the symmetries of each object. Classify each permutation of the object as a reflective symmetry or as a rotational symmetry. Write each symmetry as a permutation of the vertices (which you have already labeled). It may help if you construct a model of each object.

Further Reading

Briggs, J., *Fractals: The Patterns of Chaos,* Simon and Schuster, New York, 1992.

Field, M., and Golubitsky, M., *Symmetry in Chaos: A Search for Pattern in Mathematics, Art, and Nature*, Oxford University Press, Oxford, 1995.

Wertenbaker, C., Nature's Patterns, *Parabola*, 24, 1999, 5–12.

4.2 Symmetry and Rigid Motions

Preliminary Activity. The object drawn below does not exhibit any nontrivial symmetry. Use this image to create a pattern or design that does exhibit both reflective and rotational symmetry. This can be done in any manner desired. After the pattern has been created, write a short paragraph that describes the methods used to create it.

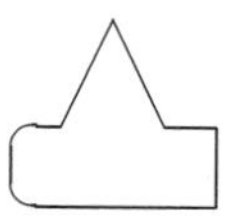

To this point, only two types of symmetries have been mentioned, reflective and rotational. In terms of planar symmetry (symmetries that exist for objects drawn on a plane), there are two more, *translational* and *glide reflective*. All four are listed next.

- **Reflective Symmetry.** An object exhibits reflective symmetry when a line or *axis of reflection* can be drawn in the plane and the portion of the object on one side of the axis is a mirror image of the portion on the other side. See Figure 4.7A.
- **Rotational Symmetry.** An object exhibits rotational symmetry when a point can be located in the plane so that when the object is rotated a certain number of degrees around that point the original object is obtained. See Figure 4.7B.
- **Translational Symmetry.** An object exhibits translational symmetry when a line or *axis of translation* can be drawn in the plane such that when one portion of the object is moved along that line a certain number of times, the entire object is obtained. See Figure 4.8A.

FIGURE 4.7
A. Reflective Symmetry B. Rotational Symmetry

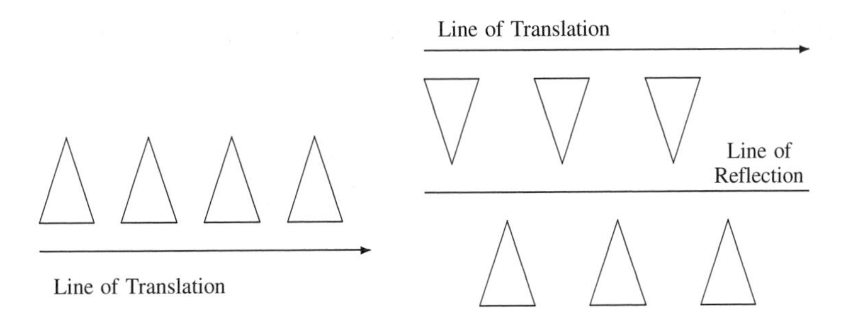

FIGURE 4.8
A. Translational Symmetry B. Glide Reflective Symmetry

- **Glide Reflective Symmetry.** A glide reflection is a combination of a translation and a reflection. An object exhibits glide reflective symmetry when an axis of translation and an axis of reflection, parallel to the axis of translation, can be drawn such that when one portion of the object is translated along the first axis and reflected over the second axis a certain number of times, the entire object is obtained. See Figure 4.8B.

Not only do objects, images, and patterns exhibit these four symmetries, but these symmetries can also be used to create objects that are symmetric. In other words, a base image can be taken and specific methods can be used to move it around the plane on which it is drawn to create an image that has symmetry. The movement methods come from the four symmetries mentioned above and are called *rigid motions*. There are four rigid motions for objects in the plane.

- **Reflection.** A reflective rigid motion is obtained by taking a base object and reflecting it over an axis of reflection. This can be done once or any number of times to generate a picture that is symmetric. See Figure 4.9.

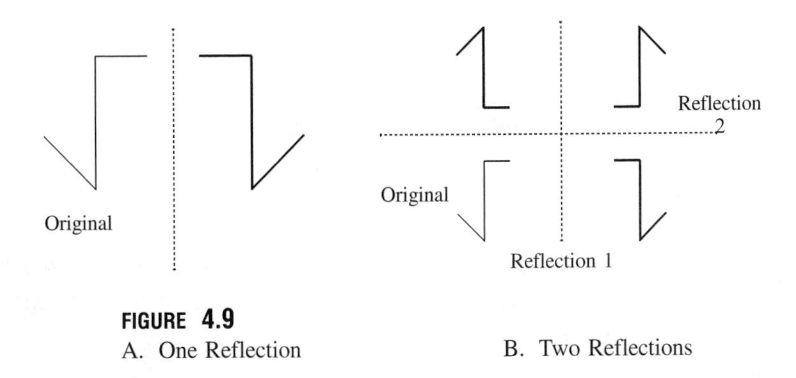

FIGURE 4.9
A. One Reflection B. Two Reflections

- **Rotation.** A rotational rigid motion is obtained when a base object is rotated around a point a certain number of times. The point around which the rotation occurs can be in the interior of the object, on the edge of the object, or somewhere outside the object. See Figure 4.10.

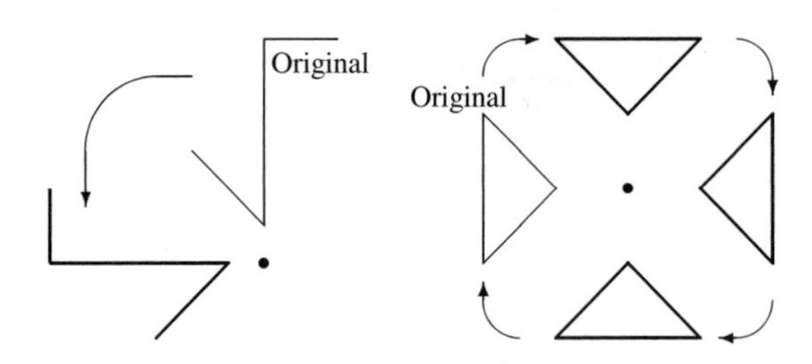

FIGURE 4.10
Examples of Rotational Rigid Motion

- **Translation.** A translational rigid motion is obtained by taking a base object and moving it along an axis of translation a certain number of times. The object can be moved once, twice, or any number of times. See Figure 4.11.

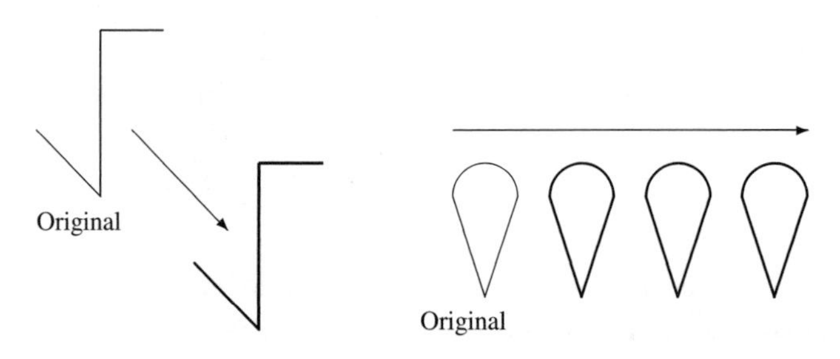

FIGURE 4.11
Examples of Translational Rigid Motions

- **Glide Reflection.** A glide reflective rigid motion is obtained by taking a base object, moving it along an axis of translation, and then reflecting it over an axis of reflection. The axes of translation and reflection must be parallel. The object can be moved any number of times using this method. See Figure 4.12.

These four rigid motions can be combined in any number of ways to create a picture that is symmetric. This combination can occur a finite number of times or an infinite number of times to fill up the entire plane.

When an infinite number of rigid motions is applied to a base object to fill the plane, a *pattern* is created.

Patterns can be created by using any number and combination of rigid motions. First the base object is picked, then copies of that object are created. The copying is done by

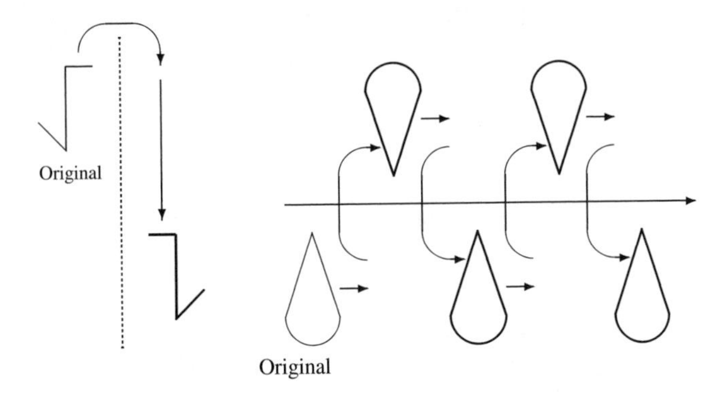

FIGURE 4.12
Examples of Glide Reflective Rigid Motion

means of rigid motions or combinations of rigid motions. The process is repeated until the plane is filled. Wallpaper patterns and floor tilings are two familiar examples of this procedure.

Figure 4.13 illustrates three kinds of pattern generation. We will examine each in turn.

- **Pattern 1.** To generate this pattern by using rigid motions, start with the rectangle in the upper left-hand corner. This rectangle is the base object of the pattern: starting with it, the entire pattern will be generated. To create the top row of the pattern, translate the base rectangle horizontally, moving it each time just enough to leave no space between rectangles. Then take the top row just created and translate it diagonally to obtain the row below it, moving it just enough to leave no space between rows. By taking these top two rows and translating them vertically, with no space between copies, the entire pattern is created.
- **Pattern 2.** Starting with the triangle in the upper left-hand corner, the base object for this design, the pattern is generated by using rigid motions. Performing a glide reflection will create the upside-down triangle just below and to the right of the base

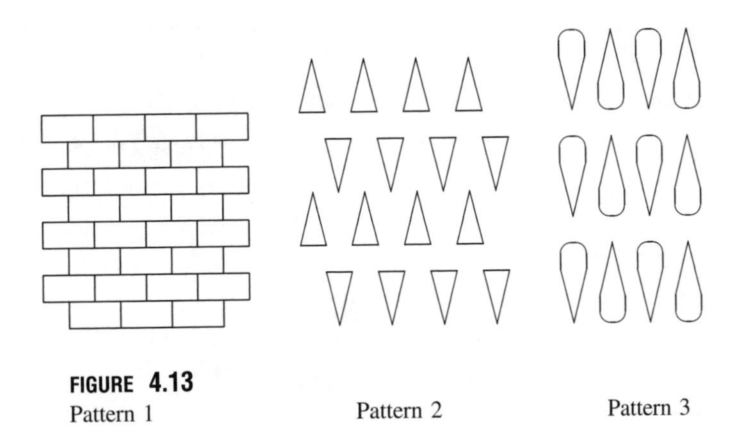

FIGURE 4.13
Pattern 1 Pattern 2 Pattern 3

FIGURE 4.14
Examples of Borders

triangle. To do this, reflect the base triangle over a horizontal line just below it and then translate it horizontally to the right. A simple horizontal translation of these two triangles will result in the top two rows of the pattern. Taking these top two rows and translating them vertically creates the entire pattern.

- **Pattern 3.** To generate this pattern through the use of rigid motions, start with the cone in the upper left-hand corner. This cone is the base object of the pattern. To create the upside down cone just to the right of the base cone, rotate the base cone 180° over a point just to the right of and halfway between the top and bottom of the base cone. By translating these two cones horizontally, the top row is obtained. Taking this top row and translating it vertically creates the entire pattern.

When a design displays only one translational axis, regardless of what other rigid motions are involved, the pattern is referred to as a *border*. Figure 4.14 shows three examples of borders.

Exercises 4.2

1. Find the different types of symmetry exhibited in each of the following diagrams. Be sure to list the symmetries and show how the diagrams exhibit those symmetries.

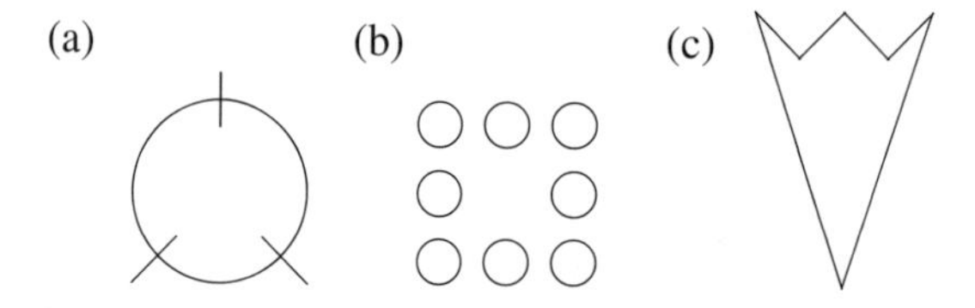

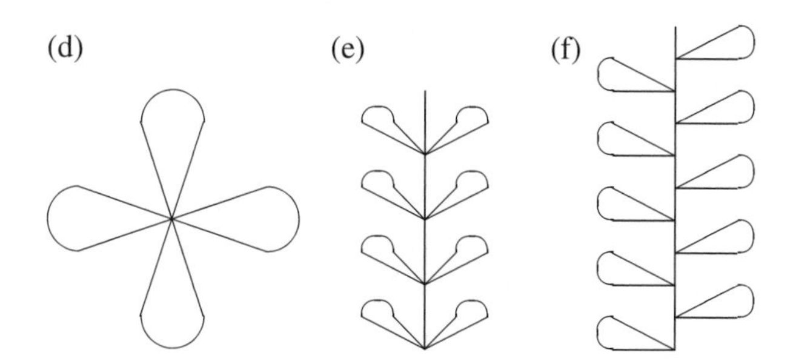

2. For each pattern illustrated below, locate a base image and show how rigid motions were used to generate the pattern (one, two, three, or all four rigid motions may be used). Be sure to list all the rigid motions involved.

   Show how each rigid motion is used to generate the diagram (draw all axes of reflection, translation, glide reflection, and points of rotation). Explain why those are the only rigid motions needed.

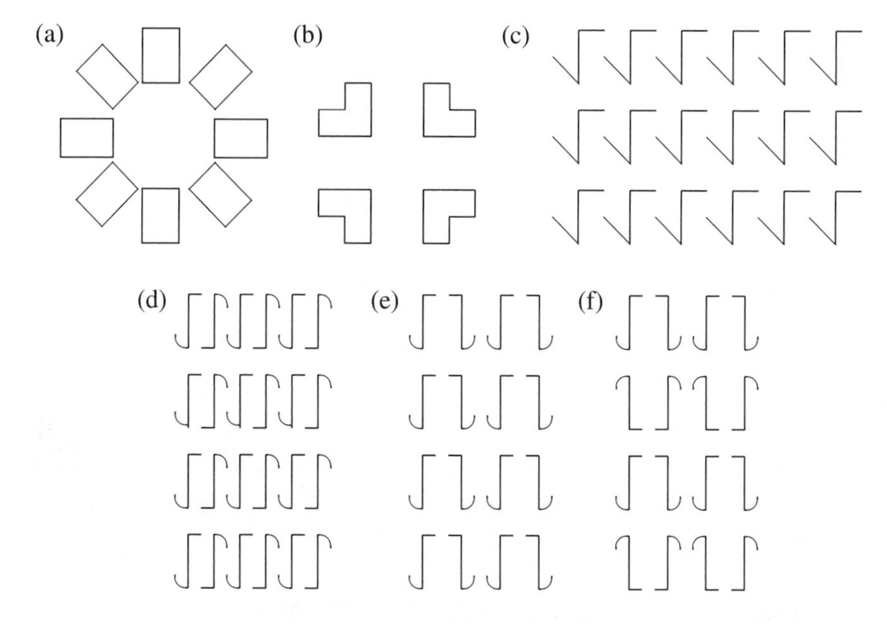

3. For each of the following images, generate a pattern by using the rigid motion(s) listed.

   (a) Translation along a horizontal axis.

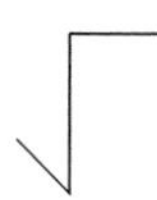

   (b) Translation along a horizontal axis and a vertical axis.

(c) Five rotations of 60°, 120°, 180°, 240°, and 300° around a point outside the triangle.

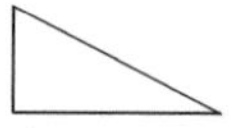

(d) Five rotations of 60°, 120°, 180°, 240°, and 300° around a point outside the triangle, plus a horizontal translation.

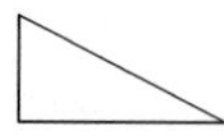

(e) Five rotations of 60°, 120°, 180°, 240°, and 300° around a point outside the triangle, a horizontal translation, and a vertical translation.

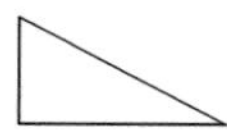

(f) Reflection over a horizontal axis below the diagram, and a horizontal translation along the same axis.

(g) Glide reflection where the axis of reflection and translation is a vertical line just to the right of the diagram.

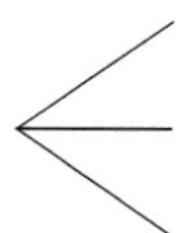

4. Take two images and create two patterns, one based on each image. To create each pattern, take the image chosen and apply any number of rigid motions to that object to fill the plane (a single sheet of paper). Clearly indicate the image and rigid motions used, and how the rigid motions were applied to the base image to generate the pattern.
5. Take two images and create two borders, one based on each image. To create each border, take the image chosen and apply any number of rigid motions. Recall that for borders, only one axis of translation is used. Clearly indicate the image and rigid motions used, and how to apply the rigid motions to the base image to generate the pattern.

Paper Assignments 4.2

1. **Description.** Pick two patterns that exhibit translational and glide reflective symmetry (e.g., wallpaper, borders, fabric design). In a short essay, discuss which pattern you prefer and why, by comparing and contrasting the patterns. You will need to describe

how each pattern is generated. Include any diagrams that will help illustrate the points that you are making in your paper.

2. **Process Description.** Start with a base object and manipulate it, using the four rigid motions of the plane, to create a pattern. The pattern created must involve at least three of the four rigid motions. In this essay, identify the base object, the rigid motions used to generate the pattern, and the order in which they were applied to generate the pattern. Include a copy of the pattern with your paper.
3. **Analysis.** The artist M. C. Escher created elaborate patterns by using the four rigid motions of the plane. What is more amazing is that the patterns he created were woven together with no holes or gaps between the copies of the base image he used. Write an essay that discusses an M.C. Escher pattern. The object of the paper is to identify the base figure or object, describe all of the rigid motions (symmetries) used in the pattern, and identify where they appear in the pattern. Specify the base figure or figures that were manipulated by the rigid motions to generate the pattern, and tell how the rigid motions generate the pattern. Feel free to include drawings or diagrams that will help illustrate the points being made.
4. **Analysis.** The work of M. C. Escher is described in Paper Assignment 4.2.3 above. Try creating a pattern similar to those created by Escher. Explain the procedure you followed to create the pattern, highlighting the rigid motions used. Specify the base figure or figures that were manipulated by the rigid motions to generate the pattern, and tell how the rigid motions generate the pattern. Feel free to include drawings or diagrams that will help illustrate the points being made.

Group Activity 4.2

1. Use the four rigid motions of the plane to create a pattern similar to those of M. C. Escher. Note the base figure or object, the rigid motions used, and how the rigid motions generate the pattern. Create a pattern that is as "tight" as possible; in other words, leave very little space between the manipulated figures.

Further Reading

Escher, M. C., *The Infinite World of M. C. Escher*, Abradale Press, New York, 1984.

Escher, M. C., *Escher on Escher,* H. N. Abrams, New York, 1989.

Field, M., and Golubitsky, M., *Symmetry in Chaos: A Search for Pattern in Mathematics, Art, and Nature*, Oxford University Press, Oxford, 1995.

Gardner, M., *Time Travel and Other Mathematical Bewilderments,* W. H. Freeman, New York, 1988.

Schattschneider, D., *Visions of Symmetry: Notebooks, Periodic Drawings, and Related Work of M. C. Escher*, W. H. Freeman, New York, 1990.

Stewart, I., The Art of Elegant Tiling, *Scientific American*, 281(1), 1999, 96–98.

5

Group Theory and the Verhoeff Check Digit Scheme

Recall the goal of this book as stated in Chapter 2: To develop a check digit scheme that at the very least catches all single-digit and transposition-of-adjacent-digits errors, that uses as check digits only the digits 0 through 9, and that works with identification numbers of any length. All of the mathematics introduced in the first four chapters will now be used to develop some concepts in group theory and to explore a scheme that meets these requirements. Specifically, we will apply the concepts of group definition, the Cayley table, and the power of a group element to the Verhoeff check digit scheme.

5.1 Fundamental Concepts

Preliminary Activity. So far, many different examples of sets have been discussed. The integers $\mathbb{Z} = \{\ldots, -3, -2, -1, 0, 1, 2, 3, \ldots\}$, the whole numbers $\{0, 1, 2, 3, 4, \ldots\}$, and the collection of digits $\{0, 1, 2, 3, 4, 5, 6, 7, 8, 9\}$ are all examples of sets. A new concept will now be introduced: *closed* sets.

An *operation* is a way of combining two elements from a set (e.g., subtraction, division). The set, along with the operation, is termed *closed* if using the operation to combine any pair of elements from the set results in another element from the set.

For example, the integers $\mathbb{Z} = \{\ldots, -3, -2, -1, 0, 1, 2, 3, \ldots\}$ along with the operation of subtraction is a closed set. It is closed because subtracting any two integers from one another results in another integer ($13 - 3 = 10$, $25 - (-105) = 130$, $6 - 0 = 6$, $20 - 80 = -60$, etc.).

On the other hand, the same set $\mathbb{Z}$ with the operation of division is not closed. Although dividing two integers does result in another integer occasionally ($\frac{8}{4} = 2$ and $-\frac{30}{6} = -5$), this is not true for all pairs of integers. For example, $\frac{1}{2} = 0.5$ and $-\frac{4}{5} = -0.8$. Neither 0.5 nor -0.8 is in $\mathbb{Z}$.

1. Consider the set of whole numbers $\{0, 1, 2, 3, 4, \ldots\}$. Find an operation on this set that will make it closed. Describe why the set along with this operation is closed. Find another operation on this set that will not make it closed. Describe why the set along with this operation is not closed.
2. Create examples to show that the set $\{0, 1, 2, 3, 4, 5, 6, 7, 8, 9\}$ is not closed under addition, subtraction, multiplication, or division. However, there are operations under which it is closed. Come up with a new operation under which the set $\{0, 1, 2, 3, 4, 5, 6, 7, 8, 9\}$ is closed. Explain why your method works.

Recall the four rigid motions of the plane that were introduced in Chapter 4. Each one—reflection, rotation, translation, and glide reflection—is a way to move an object in the plane to achieve a picture that is symmetric or that displays symmetry. These rigid motions can be used to create elaborate designs, patterns, and borders. The four rigid motions also have an interesting property. If any two rigid motions are applied to an object, one after the other, the result of this combination can always be obtained by another single rigid motion.

Example 5.1.1. *The simplest example is a reflection followed by a translation where the axes of reflection and translation are parallel. By definition, the combination of these two rigid motions is the glide reflective rigid motion.*

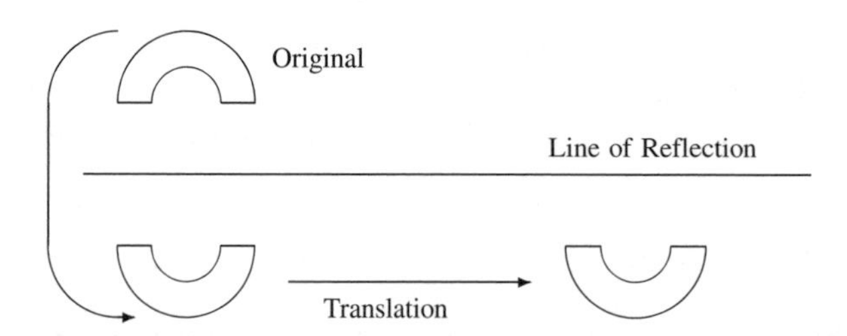

Example 5.1.2. *The movement obtained by performing a reflection over a horizontal line followed by a downward diagonal translation is also attainable through a single rotation.*

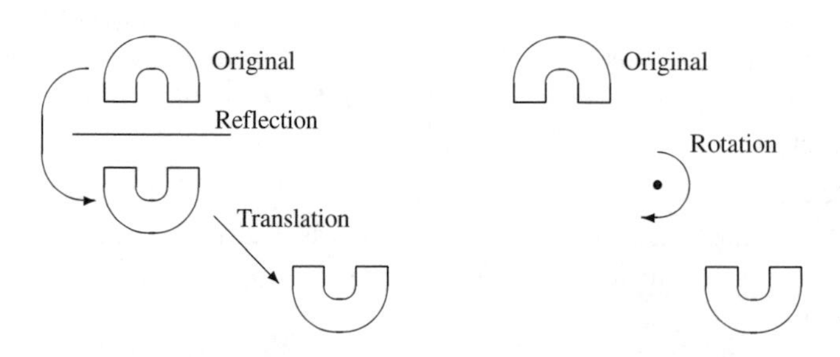

Example 5.1.3. *Combining a* $90°$ *rotation with a horizontal translation is the same as completing a single* $90°$ *rotation over a different point.*

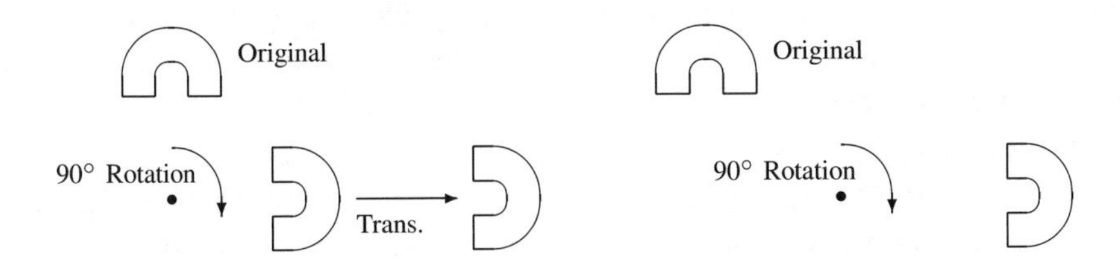

Example 5.1.4. *Combining a glide reflection with a rotation can be obtained through a different single glide reflection.*

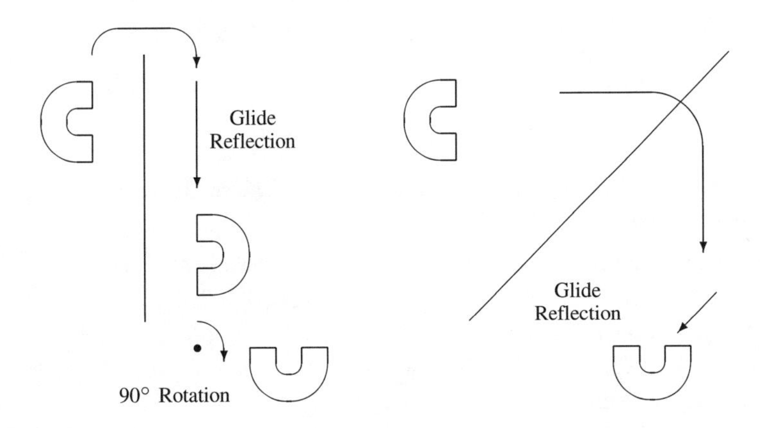

The combinations presented in the preceding examples demonstrate the following principle: When any two elements from the collection of all rigid motions are combined, another rigid motion or element from that set is obtained. In other words, the collection of rigid motions forms a *closed* set.

A method of combining the elements of a set is called an *operation.* Given a set and an operation combining any two of its elements, that set is closed if the operation results in an element that is also in the set.

There are many examples of closed sets. The set of integers

$$\mathbb{Z} = \{\dots, -3, -2, -1, 0, 1, 2, 3, \dots\}$$

along with the operation of subtraction is closed. This was demonstrated in the Preliminary Activity at the beginning of this section. In that activity, it was also shown that the same set $\mathbb{Z}$ with the operation of division is not closed.

Another example of a closed set is the set of non-negative integers $\{0, 1, 2, 3, 4, \ldots\}$ along with the operation of addition. This set is closed because any two non-negative integers added together always result in a non-negative integer. For example, $3+4=7$, $9+0=9$, $18+18=36$, and so on.

However, the set of digits $\{0,1,2,3,4,5,6,7,8,9\}$ with the operation of addition is *not* closed. Even though some of the numbers from this set can be added together to give another number from the set ($1+7=8$, $3+3=6$, etc.), this is not true for all pairs of numbers. For example, $6+9=15$, and 15 is not in the set.

Although the set $\{0,1,2,3,4,5,6,7,8,9\}$ is not closed under addition, there is an operation under which it is closed. If we add any two integers (a and b) from this set, divide this sum by 10, and find the remainder—in other words, if we calculate $(a+b) \pmod{10}$—the result will always be an element of this set. Recall that the only possible remainders when any number is divided by 10 are $0,1,2,3,4,5,6,7,8$, or 9. For example,

$$\begin{aligned}(1+7) \pmod{10} &= 8 \pmod{10} = 8,\\ (3+3) \pmod{10} &= 6 \pmod{10} = 6,\\ (6+9) \pmod{10} &= 15 \pmod{10} = 5, \quad \text{and}\\ (9+9) \pmod{10} &= 18 \pmod{10} = 8.\end{aligned}$$

Thus the set $\{0,1,2,3,4,5,6,7,8,9\}$ with the operation $\oplus_{10}$, defined by

$$a \oplus_{10} b = (a+b) \pmod{10},$$

is closed. The symbol $\oplus_{10}$ denotes addition modulo 10. The set $\{0,1,2,3,4,5,6,7,8,9\}$ along with the operation $\oplus_{10}$ is denoted by $\mathbb{Z}_{10}$.

There is nothing unique about modulo 10 arithmetic. Consider the operation of addition modulo n, denoted by the symbol $\oplus_n$. The set of integers $\{0,1,2,3,\ldots,n-1\}$ along with the operation $\oplus_n$, where for a and $b \in \{0,1,2,3,\ldots,n-1\}$, $a \oplus_n b = (a+b) \pmod{n}$, is always a closed set for any $n \geq 1$. The set $\{0,1,2,\ldots,n-1\}$ with the operation $\oplus_n$ is denoted by $\mathbb{Z}_n$. For example, $\mathbb{Z}_{15}$ is the set $\{0,1,2,3,4,5,6,7,8,9,10,11,12,13,14\}$ with the operation $\oplus_{15}$. Here,

$$\begin{aligned}8 \oplus_{15} 4 &= (8+4) \pmod{15} = 12 \pmod{15} = 12 \quad \text{and}\\ 12 \oplus_{15} 14 &= (12+14) \pmod{15} = 26 \pmod{15} = 11,\end{aligned}$$

since $26 = 1 \cdot 15 + 11$.

Another example of a closed set is D_{10}. If any two symmetries of the pentagon are combined, the result is another symmetry of the pentagon. Consider these two examples.

Example 5.1.5. *When rf_3 is followed by rt_2, the symmetry rf_4 results, as shown.*

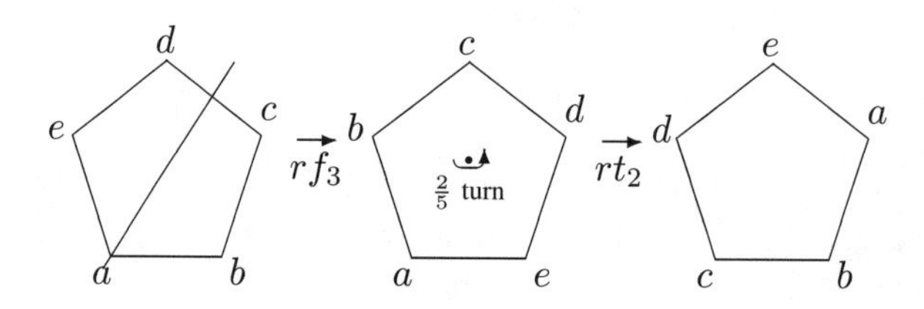

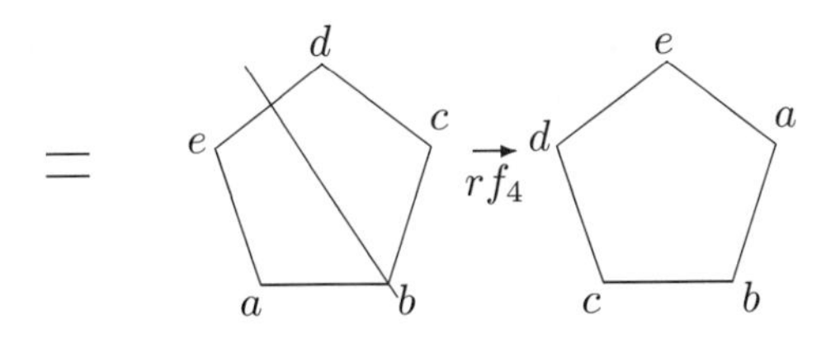

Example 5.1.6. *The symmetry rt_1 followed by rt_3 equals the symmetry rt_4, as shown.*

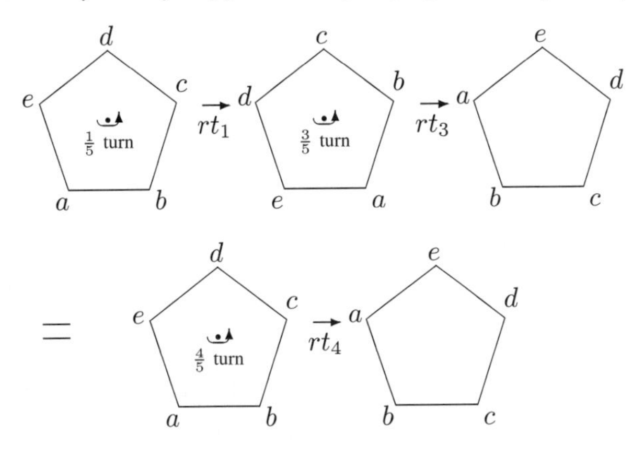

Two other important mathematical concepts, *identity* and *inverse*, can be illustrated by the following simple exercise from algebra: Solve the equation $x+5=14$ for x. To find x, a technique similar to the method presented below would be used:

$$
\begin{aligned}
x+5 &= 14 \\
x+5+(-5) &= 14+(-5) \qquad \text{(add } -5 \text{ to both sides)} \\
x+5-5 &= 14-5 \\
x+0 &= 9 \\
x &= 9
\end{aligned}
$$

The first step was to isolate the x by adding -5 to both sides of the equation. This was done because $5+(-5)=5-5=0$. This resulted in the equation $x+0=9$. Since $x+0=x$, the answer is $x=9$. Changing $x+5$ into $x+0$ was the big step in solving this equation. This could be done since $a+0=a$ for any number a.

Given this special property of zero, 0 is sometimes referred to as an *identity*. In other words, adding 0 to any number does not change the value of that number. It remains identically the same. This concept is similar to the identity symmetry id of the pentagon. Applying this symmetry to the pentagon does not move (reflect or rotate) it.

To get 0, -5 was added to both sides of the equation. This was done to change the $x+5$ into $x+0$. In a sense, -5 is the opposite of 5 and when added to 5 gives the identity 0. The integer -5 is referred to as the *additive inverse* of 5. Under addition, any integer n has an inverse $-n$. The inverse of 18 is -18, as $18+(-18)=18-18=0$. The inverse of -7 is $-(-7)=7$, as $-7+7=0$. This inverse concept can also be seen in the symmetries of the pentagon, as the following example illustrates.

Example 5.1.7. *The symmetry rt_4 is the inverse of rt_1 since applying rt_4 after rt_1 results in the identity symmetry id, as shown.*

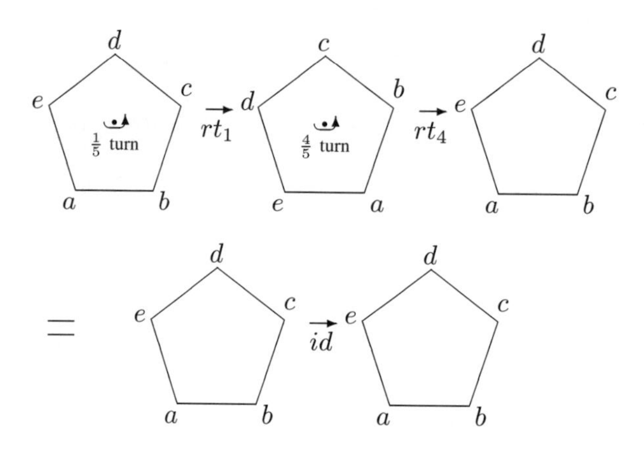

The concepts of an identity and an inverse are not unique to the integers under addition. A variety of sets and operations exhibit the same properties. The operation associated with a generic set is denoted by an asterisk $*$.

Definition 5.1.8. *If a non-empty set G has an operation $*$ associated with it, G is called a* ***group*** *when the following properties hold:*

1. *Closure:* For any two elements a and b in G, $a * b$ is also in G.
2. *Associative:* For all elements a, b, and c in G, $(a * b) * c = a * (b * c)$.
3. *Identity:* G contains an element e, called the **identity** of G, such that $e*a = a*e = a$.
4. *Inverse:* For each element a in G, there is an element a^{-1}, called the **inverse** of a, such that $a * a^{-1} = a^{-1} * a = e$.

If a group G is has a finite number of elements, G is called a *finite group*. If it has an infinite number of elements, it is called an *infinite group*. Examples of finite groups include $\mathbb{Z}_n$, S_n, and D_{2n}. Each one will be examined in detail to show how they satisfy the definition of a group.

Example 5.1.9. *The set $\mathbb{Z}_n = \{0, 1, 2, 3, \ldots, n-1\}$ with the operation $\oplus_n$, where $a \oplus_n b = (a+b) \pmod{n}$ for all elements a and b in $\mathbb{Z}_n$, satisfies the four criteria from Definition 5.1.8, so it is a group.*

1. *Closure.* The operation $\oplus_n$ satisfies the first criterion from the definition (it is closed), since any number divided by n has as a possible remainder only $0, 1, 2, 3, \ldots,$ or $n-1$. These digits are exactly those in $\mathbb{Z}_n$.

   For example, consider $\mathbb{Z}_{10} = \{0, 1, 2, 3, 4, 5, 6, 7, 8, 9\}$ with the operation $\oplus_{10}$. When the sum of any two digits is divided by 10, the remainders will always be

$0, 1, 2, 3, 4, 5, 6, 7, 8,$ or 9. For instance,

$$\begin{aligned} 5 \oplus_{10} 4 &= (5+4) \pmod{10} = 9 \pmod{10} = 9; \\ 7 \oplus_{10} 6 &= (7+6) \pmod{10} = 13 \pmod{10} = 3; \quad \text{and} \\ 8 \oplus_{10} 9 &= (8+9) \pmod{10} = 17 \pmod{10} = 7. \end{aligned}$$

2. *Associative.* The operation $\oplus_n$ satisfies the second criterion (it is associative), since regular addition $(+)$ is itself associative. Consider the following example from $\mathbb{Z}_{10} = \{0, 1, 2, 3, 4, 5, 6, 7, 8, 9\}$ with the operation $\oplus_{10}$ and $a = 6$, $b = 4$, and $c = 9$. The computations result in

$$\begin{aligned} (a \oplus_{10} b) \oplus_{10} c &= (6 \oplus_{10} 4) \oplus_{10} 9 \\ &= \big((6+4) \pmod{10}\big) \oplus_{10} 9 \\ &= \big(10 \pmod{10}\big) \oplus_{10} 9 \\ &= 0 \oplus_{10} 9 \\ &= (0+9) \pmod{10} \\ &= 9 \pmod{10} \\ &= 9 \end{aligned}$$

and

$$\begin{aligned} a \oplus_{10} (b \oplus_{10} c) &= 6 \oplus_{10} (4 \oplus_{10} 9) \\ &= 6 \oplus_{10} \big((4+9) \pmod{10}\big) \\ &= 6 \oplus_{10} \big(13 \pmod{10}\big) \\ &= 6 \oplus_{10} 3 \\ &= (6+3) \pmod{10} \\ &= 9 \pmod{10} \\ &= 9. \end{aligned}$$

3. *Identity.* The set $\mathbb{Z}_n = \{0, 1, 2, 3, \ldots, n-1\}$ with the operation $\oplus_n$ also satisfies the third criterion, because it has an identity element e. Recall that the identity element e must satisfy the property that $e \oplus_n a = a \oplus_n e = a$ for all $a \in \mathbb{Z}_n$. For $\mathbb{Z}_n$, $e = 0$. For any element $a \in \mathbb{Z}_n$,

$$0 \oplus_n a = (0 + a) \pmod{n} = a \pmod{n} = a,$$

since $0 \le a \le n-1$. A similar argument shows that $a \oplus_n 0 = a$. In our example of $\mathbb{Z}_{10} = \{0, 1, 2, 3, 4, 5, 6, 7, 8, 9\}$ with the operation $\oplus_{10}$, the identity $e = 0$. Take a few moments and prove it.

4. *Inverse.* The set $\mathbb{Z}_n = \{0, 1, 2, 3, \ldots, n-1\}$ with the operation $\oplus_n$ satisfies the fourth criterion, because every element in $\mathbb{Z}_n$ has an inverse. To show this, it must be shown that for each element $a \in \mathbb{Z}_n$, there is an element $a^{-1} \in \mathbb{Z}_n$ such that $a \oplus_n a^{-1} = a^{-1} \oplus_n a = 0$, where $0 = e$ (the identity element). It is not too hard to show that for $a \in \mathbb{Z}_n$, $a^{-1} = n - a$.

   Consider again our example of $\mathbb{Z}_{10} = \{0, 1, 2, 3, 4, 5, 6, 7, 8, 9\}$ with the operation $\oplus_{10}$. For $a = 3$, $a^{-1} = 3^{-1} = 10 - 3 = 7$. Note that

$$\begin{aligned} a \oplus_{10} a^{-1} &= 3 \oplus_n 7 = (3+7) \pmod{10} = 10 \pmod{10} = 0 \quad \text{and} \\ a^{-1} \oplus_{10} a &= 7 \oplus_n 3 = (7+3) \pmod{10} = 10 \pmod{10} = 0. \end{aligned}$$

An element can be its own inverse. Consider $b = 5$. Since $b^{-1} = 5^{-1} = 10 - 5 = 5$,

$$b \oplus_{10} b^{-1} = 5 \oplus_n 5 = (5+5) \pmod{10} = 10 \pmod{10} = 0$$

and

$$b^{-1} \oplus_{10} b = 5 \oplus_n 5 = (5+5) \pmod{10} = 10 \pmod{10} = 0.$$

Our second example of a group will be S_n, the collection of permutations of the set $\{0, 1, 2, 3, \ldots, n-1\}$. Before this can be established, however, an operation needs to be defined on S_n (recall that an operation is a method for combining two elements from a set).

The operation defined on S_n is *composition of functions* and is denoted by the small circle $\circ$. Given two permutations f and g from S_n, $f \circ g$ results in the permutation of $\{0, 1, 2, 3, \ldots, n-1\}$ that is obtained by applying first f and then g.

Consider S_4, the collection of permutations on $\{0, 1, 2, 3\}$ and the two permutations

$$f = \begin{pmatrix} 0 & 1 & 2 & 3 \\ 2 & 3 & 1 & 0 \end{pmatrix} \quad \text{and} \quad g = \begin{pmatrix} 0 & 1 & 2 & 3 \\ 0 & 3 & 1 & 2 \end{pmatrix}$$

of S_4. Applying first f and then g can be visualized in the following manner:

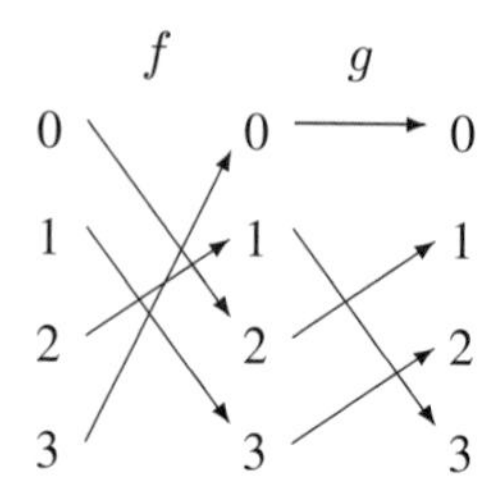

To obtain the composition of f and g, denoted by the permutation $f \circ g$, follow the paths that have been created.

- First, f sends $0 \to 2$. Follow this by g, which sends $2 \to 1$. Thus the composition $f \circ g$ sends $0 \to 1$. In other words, since the path starts at 0 and ends at 1, $f \circ g$ sends $0 \to 1$.
- Next, f sends $1 \to 3$. Follow this by g, which sends $3 \to 2$. As a result, the composition $f \circ g$ sends $1 \to 2$. In other words, since the path starts at 1 and ends at 2, $f \circ g$ sends $1 \to 2$.
- Next, f sends $2 \to 1$. Follow this by g, which sends $1 \to 3$. Consequently, the composition $f \circ g$ sends $2 \to 3$. In other words, since the path starts at 2 and ends at 3, $f \circ g$ sends $2 \to 3$.
- Finally, f sends $3 \to 0$. Follow this by g, which sends $0 \to 0$. Thus the composition $f \circ g$ sends $3 \to 0$. In other words, since the path starts at 3 and ends at 0, $f \circ g$ sends $3 \to 0$.

The new permutation $f \circ g$ of S_4 has now been created. The result, shown next, is clearly a permutation, as it is one-to-one and onto.

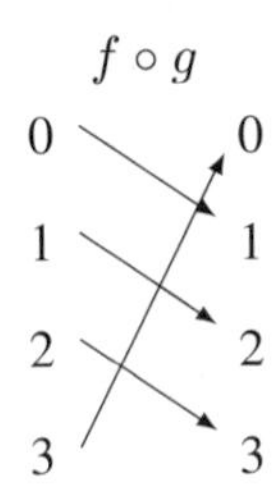

Thus $f \circ g = \left(\begin{smallmatrix} 0 & 1 & 2 & 3 \\ 1 & 2 & 3 & 0 \end{smallmatrix}\right)$ is an element of S_4. To find $f \circ g$, drawing a picture of the two permutations is not necessary. All we need are the permutation representations of f and g:

$$f \circ g = \begin{pmatrix} 0 & 1 & 2 & 3 \\ 2 & 3 & 1 & 0 \end{pmatrix} \circ \begin{pmatrix} 0 & 1 & 2 & 3 \\ 0 & 3 & 1 & 2 \end{pmatrix}.$$

To find where $f \circ g$ sends 0, follow 0 as it is first permuted to 2 by f and then follow the 2 as it is permuted by g to 1. This process is illustrated below.

$$\begin{bmatrix} 0 & 1 & 2 & 3 \\ 2 & 3 & 1 & 0 \end{bmatrix} \circ \begin{bmatrix} 0 & 1 & 2 & 3 \\ 0 & 3 & 1 & 2 \end{bmatrix}$$

Since f sends $0 \to 2$ and g sends $2 \to 1$, the composition sends $0 \to 1$. Proceeding in this manner, the new permutation

$$\begin{aligned} f \circ g &= \begin{pmatrix} 0 & 1 & 2 & 3 \\ 2 & 3 & 1 & 0 \end{pmatrix} \circ \begin{pmatrix} 0 & 1 & 2 & 3 \\ 0 & 3 & 1 & 2 \end{pmatrix} \\ &= \begin{pmatrix} 0 & 1 & 2 & 3 \\ 1 & 2 & 3 & 0 \end{pmatrix} \end{aligned}$$

is obtained. Note that order is very important. Although $f \circ g = \left(\begin{smallmatrix} 0 & 1 & 2 & 3 \\ 1 & 2 & 3 & 0 \end{smallmatrix}\right)$, the composition $g \circ f = \left(\begin{smallmatrix} 0 & 1 & 2 & 3 \\ 2 & 0 & 3 & 1 \end{smallmatrix}\right)$ results in a very different permutation in S_4.

No matter what the value of n is, composition of two permutations in S_n is defined the same way. Work out the following examples to convince yourself of this fact.

- For $f = \left(\begin{smallmatrix} 0 & 1 & 2 & 3 \\ 3 & 2 & 1 & 0 \end{smallmatrix}\right)$ and $g = \left(\begin{smallmatrix} 0 & 1 & 2 & 3 \\ 1 & 0 & 2 & 3 \end{smallmatrix}\right)$ in S_4,

$$\begin{aligned} f \circ g &= \begin{pmatrix} 0 & 1 & 2 & 3 \\ 3 & 2 & 1 & 0 \end{pmatrix} \circ \begin{pmatrix} 0 & 1 & 2 & 3 \\ 1 & 0 & 2 & 3 \end{pmatrix} \\ &= \begin{pmatrix} 0 & 1 & 2 & 3 \\ 3 & 2 & 0 & 1 \end{pmatrix}. \end{aligned}$$

- For $f = \left(\begin{smallmatrix} 0 & 1 & 2 & 3 & 4 & 5 & 6 \\ 1 & 0 & 6 & 5 & 3 & 4 & 2 \end{smallmatrix}\right)$ and $g = \left(\begin{smallmatrix} 0 & 1 & 2 & 3 & 4 & 5 & 6 \\ 4 & 3 & 1 & 2 & 0 & 6 & 5 \end{smallmatrix}\right)$ in S_7,

$$\begin{aligned} f \circ g &= \begin{pmatrix} 0 & 1 & 2 & 3 & 4 & 5 & 6 \\ 1 & 0 & 6 & 5 & 3 & 4 & 2 \end{pmatrix} \circ \begin{pmatrix} 0 & 1 & 2 & 3 & 4 & 5 & 6 \\ 4 & 3 & 1 & 2 & 0 & 6 & 5 \end{pmatrix} \\ &= \begin{pmatrix} 0 & 1 & 2 & 3 & 4 & 5 & 6 \\ 3 & 4 & 5 & 6 & 2 & 0 & 1 \end{pmatrix}. \end{aligned}$$

- For $f = \left(\begin{smallmatrix} 0&1&2&3&4&5&6&7&8&9 \\ 2&3&4&9&8&7&0&6&5&1 \end{smallmatrix}\right)$ and $g = \left(\begin{smallmatrix} 0&1&2&3&4&5&6&7&8&9 \\ 4&6&7&3&5&9&2&1&8&0 \end{smallmatrix}\right)$ in S_{10},

$$f \circ g = \begin{pmatrix} 0&1&2&3&4&5&6&7&8&9 \\ 2&3&4&9&8&7&0&6&5&1 \end{pmatrix} \circ \begin{pmatrix} 0&1&2&3&4&5&6&7&8&9 \\ 4&6&7&3&5&9&2&1&8&0 \end{pmatrix}$$
$$= \begin{pmatrix} 0&1&2&3&4&5&6&7&8&9 \\ 7&3&5&0&8&1&4&2&9&6 \end{pmatrix}.$$

The operation of composition of functions, denoted by $\circ$, and the fact that it is closed (i.e., two permutations from S_n combined by $\circ$ result in another permutation in S_n) has now been established. We continue with our second example of a group.

Example 5.1.10. *S_n, the collection of permutations of the set $\{0, 1, 2, 3, \ldots, n-1\}$, is a group. We have already shown that $\circ$ is closed. It must further be shown that $\circ$ is associative, that there is an identity element e in S_n, and that each element has an inverse.*

1. *Closure.* The operation of composition of functions, denoted by $\circ$, is closed. That is, two permutations from S_n combined by $\circ$ result in another permutation in S_n, as demonstrated above.
2. *Associative.* It is a lengthy process to show that $\circ$ is associative, but a simple example from S_4 will demonstrate that it is. Recall, $\circ$ is associative when $(f \circ g) \circ h = f \circ (g \circ h)$ for all f, g, and h in S_4. Here, let $f = \left(\begin{smallmatrix} 0&1&2&3 \\ 0&3&2&1 \end{smallmatrix}\right)$, $g = \left(\begin{smallmatrix} 0&1&2&3 \\ 1&0&3&2 \end{smallmatrix}\right)$, and $h = \left(\begin{smallmatrix} 0&1&2&3 \\ 3&1&2&0 \end{smallmatrix}\right)$. The computations result in

$$(f \circ g) \circ h = \left(\begin{pmatrix} 0&1&2&3 \\ 0&3&2&1 \end{pmatrix} \circ \begin{pmatrix} 0&1&2&3 \\ 1&0&3&2 \end{pmatrix} \right) \circ \begin{pmatrix} 0&1&2&3 \\ 3&1&2&0 \end{pmatrix}$$
$$= \begin{pmatrix} 0&1&2&3 \\ 1&2&3&0 \end{pmatrix} \circ \begin{pmatrix} 0&1&2&3 \\ 3&1&2&0 \end{pmatrix}$$
$$= \begin{pmatrix} 0&1&2&3 \\ 1&2&0&3 \end{pmatrix} \qquad \text{and}$$

$$f \circ (g \circ h) = \begin{pmatrix} 0&1&2&3 \\ 0&3&2&1 \end{pmatrix} \circ \left(\begin{pmatrix} 0&1&2&3 \\ 1&0&3&2 \end{pmatrix} \circ \begin{pmatrix} 0&1&2&3 \\ 3&1&2&0 \end{pmatrix} \right)$$
$$= \begin{pmatrix} 0&1&2&3 \\ 0&3&2&1 \end{pmatrix} \circ \begin{pmatrix} 0&1&2&3 \\ 1&3&0&2 \end{pmatrix}$$
$$= \begin{pmatrix} 0&1&2&3 \\ 1&2&0&3 \end{pmatrix}.$$

3. *Identity.* The third criterion for S_n to be a group is an identity element $e \in S_n$ such that $f \circ e = e \circ f = f$ for all $f \in S_n$. Let $f = \left(\begin{smallmatrix} 0 & 1 & 2 & \ldots & n-1 \\ f(0) & f(1) & f(2) & \ldots & f(n-1) \end{smallmatrix}\right)$ be an arbitrary permutation in S_n.

   - Since f sends $0 \to f(0)$ and $f \circ e = f$, e must send $f(0) \to f(0)$.
   - Since f sends $1 \to f(1)$ and $f \circ e = f$, e must send $f(1) \to f(1)$, and so on.

   In other words, e must be the permutation in S_n that sends each number from the set $\{0, 1, 2, \ldots, n-1\}$ to itself. This can also be written as $e = \left(\begin{smallmatrix} 0&1&2&\ldots&n-1 \\ 0&1&2&\ldots&n-1 \end{smallmatrix}\right)$. For example, we can write that

$$\text{in } S_4, \quad e = \begin{pmatrix} 0 & 1 & 2 & 3 \\ 0 & 1 & 2 & 3 \end{pmatrix};$$

$$\text{in } S_7, \quad e = \begin{pmatrix} 0 & 1 & 2 & 3 & 4 & 5 & 6 \\ 0 & 1 & 2 & 3 & 4 & 5 & 6 \end{pmatrix};$$

$$\text{in } S_{10}, \quad e = \begin{pmatrix} 0 & 1 & 2 & 3 & 4 & 5 & 6 & 7 & 8 & 9 \\ 0 & 1 & 2 & 3 & 4 & 5 & 6 & 7 & 8 & 9 \end{pmatrix}.$$

4. *Inverse.* Finally, for S_n to be a group, it must satisfy the fourth criterion: For each element $f \in S_n$, there must be an element $f^{-1} \in S_n$ such that $f \circ f^{-1} = f^{-1} \circ f = e$. Given a permutation f, f^{-1} is easy to find. In a sense, f^{-1} undoes what f does.

   Let's look at the permutation $f = \left(\begin{smallmatrix} 0 & 1 & 2 & 3 \\ 1 & 3 & 0 & 2 \end{smallmatrix}\right)$ from S_4. Whatever f^{-1} is, it must satisfy $f \circ f^{-1} = e$. At the moment, since it is not known what f^{-1} is, we can let $f^{-1} = \left(\begin{smallmatrix} 0 & 1 & 2 & 3 \\ a_0 & a_1 & a_2 & a_3 \end{smallmatrix}\right)$ where a_0, a_1, a_2, and a_3 are distinct elements of the set $\{0, 1, 2, 3\}$. The permutation f^{-1} must satisfy the equation

$$f \circ f^{-1} = e$$

or

$$\begin{pmatrix} 0 & 1 & 2 & 3 \\ 1 & 3 & 0 & 2 \end{pmatrix} \circ \begin{pmatrix} 0 & 1 & 2 & 3 \\ a_0 & a_1 & a_2 & a_3 \end{pmatrix} = \begin{pmatrix} 0 & 1 & 2 & 3 \\ 0 & 1 & 2 & 3 \end{pmatrix}.$$

Visually, this can be represented as follows:

$$\begin{array}{ccccccc} & f & \circ & f^{-1} & = & & e \\ 0 & & 0 & 0 & & 0 & \longrightarrow 0 \\ 1 & & 1 & 1 & & 1 & \longrightarrow 1 \\ 2 & & 2 & 2 & & 2 & \longrightarrow 2 \\ 3 & & 3 & 3 & & 3 & \longrightarrow 3 \end{array}$$

(with arrows for f: $0 \to 1$, $1 \to 3$, $2 \to 0$, $3 \to 2$)

- To find f^{-1}, first determine where $f \circ f^{-1}$ sends 0: $0 \xrightarrow{f} 1 \xrightarrow{f^{-1}} a_1$. Recall that $f \circ f^{-1} = e$ is needed; or, in this case, that $f \circ f^{-1}$ sends $0 \to 0$. Since $f \circ f^{-1}$ sends $0 \to a_1$, a_1 must equal 0. Thus $a_1 = 0$.

- Second, determine where $f \circ f^{-1}$ sends 1: $1 \xrightarrow{f} 3 \xrightarrow{f^{-1}} a_3$. Recall that $f \circ f^{-1} = e$ is needed; or, in this case, that $f \circ f^{-1}$ sends $1 \to 1$. Since $f \circ f^{-1}$ sends $1 \to a_3$, a_3 must equal 1. Thus $a_3 = 1$.

- Moving on to 2, $2 \xrightarrow{f} 0 \xrightarrow{f^{-1}} a_0$. Recall that $f \circ f^{-1} = e$ is required; or, in this case, that $f \circ f^{-1}$ sends $2 \to 2$. Since $f \circ f^{-1}$ sends $2 \to a_0$, a_0 must equal 2. Thus $a_0 = 2$.

- Finishing with 3, $3 \xrightarrow{f} 2 \xrightarrow{f^{-1}} a_2$. Recall that $f \circ f^{-1} = e$ is required; or, in this case, that $f \circ f^{-1}$ sends $3 \to 3$. Since $f \circ f^{-1}$ sends $3 \to a_2$, a_2 must equal 3. Thus $a_2 = 3$.

Consequently, $f^{-1} = \left(\begin{smallmatrix} 0 & 1 & 2 & 3 \\ 2 & 0 & 3 & 1 \end{smallmatrix}\right)$. It is easy to check that $f^{-1} \circ f = e$ is also true. It is also evident that f^{-1} is a member of S_4 as well.

For practice, given $g = \left(\begin{smallmatrix} 0 & 1 & 2 & 3 \\ 1 & 2 & 3 & 0 \end{smallmatrix}\right)$ and $h = \left(\begin{smallmatrix} 0 & 1 & 2 & 3 \\ 3 & 0 & 1 & 2 \end{smallmatrix}\right)$ from S_4, find g^{-1} and h^{-1}. Note that work done can always be checked, since $g \circ g^{-1} = g^{-1} \circ g = e$ and $h \circ h^{-1} = h^{-1} \circ h = e$. You should obtain the permutations $g^{-1} = \left(\begin{smallmatrix} 0 & 1 & 2 & 3 \\ 3 & 0 & 1 & 2 \end{smallmatrix}\right)$ and $h^{-1} = \left(\begin{smallmatrix} 0 & 1 & 2 & 3 \\ 1 & 2 & 3 & 0 \end{smallmatrix}\right)$.

The same technique works for any value of n in S_n. Find the inverse of each of elements $f = \left(\begin{smallmatrix} 0 & 1 & 2 & 3 & 4 & 5 \\ 1 & 2 & 4 & 5 & 0 & 3 \end{smallmatrix}\right)$ and $g = \left(\begin{smallmatrix} 0 & 1 & 2 & 3 & 4 & 5 \\ 0 & 5 & 3 & 2 & 1 & 4 \end{smallmatrix}\right)$ from S_6. The permutations $f^{-1} = \left(\begin{smallmatrix} 0 & 1 & 2 & 3 & 4 & 5 \\ 4 & 0 & 1 & 5 & 2 & 3 \end{smallmatrix}\right)$ and $g^{-1} = \left(\begin{smallmatrix} 0 & 1 & 2 & 3 & 4 & 5 \\ 0 & 4 & 3 & 2 & 5 & 1 \end{smallmatrix}\right)$ should be obtained.

Thus it has been demonstrated that S_n, for $n \geq 1$, is a group.

Example 5.1.11. *The dihedral group D_{2n} is also a group, as the name indicates.*

1. Recall that D_{2n} is the collection of symmetries of a regular n-gon. In particular, the group D_{10} is the collection of symmetries of a regular 5-gon (a pentagon). In Chapter 4, each symmetry of D_{10} was expressed in permutation notation.

   Earlier in this section, it was noted that combining any two symmetries in D_{10} resulted in another symmetry in D_{10}. This method of combining symmetries is just another way of expressing a composition of two permutations, as was done with the permutations in S_n. Recall that an element in D_{2n}, or a symmetry of a regular n-gon, can be expressed as a permutation of the vertices a, b, c, and so on. Consequently, combining symmetries from D_{2n} works the same way as combining permutations from S_n.

   For instance, it was shown that combining rf_3 with rt_2 resulted in rf_4. In the second example, combining rt_1, followed by rt_3, gave the symmetry rt_4. Using the permutation notation for each symmetry, these combinations can be expressed as

$$rf_3 \circ rt_2 = \begin{pmatrix} a & b & c & d & e \\ a & e & d & c & b \end{pmatrix} \circ \begin{pmatrix} a & b & c & d & e \\ c & d & e & a & b \end{pmatrix} = \begin{pmatrix} a & b & c & d & e \\ c & b & a & e & d \end{pmatrix} = rf_4$$

   and

$$rt_1 \circ rt_3 = \begin{pmatrix} a & b & c & d & e \\ b & c & d & e & a \end{pmatrix} \circ \begin{pmatrix} a & b & c & d & e \\ d & e & a & b & c \end{pmatrix} = \begin{pmatrix} a & b & c & d & e \\ e & a & b & c & d \end{pmatrix} = rt_4.$$

2. All the details showing that D_{2n} with the operation $\circ$ forms a group will not be presented. The identity element from D_{2n} is the symmetry that sends every vertex to itself, and finding inverses works the same way that it did in S_n.

In D_{10}, the identity element $e = \left(\begin{smallmatrix} a & b & c & d & e \\ a & b & c & d & e \end{smallmatrix}\right) = id$. The inverse of $rt_2 = \left(\begin{smallmatrix} a & b & c & d & e \\ c & d & e & a & b \end{smallmatrix}\right)$ is

$$rt_2^{-1} = \begin{pmatrix} a & b & c & d & e \\ d & e & a & b & c \end{pmatrix} = rt_3$$

and the inverse of $rf_3 = \left(\begin{smallmatrix} a & b & c & d & e \\ a & e & d & c & b \end{smallmatrix}\right)$ is

$$rf_3^{-1} = \begin{pmatrix} a & b & c & d & e \\ a & e & d & c & b \end{pmatrix} = rf_3 \quad \text{(itself)}.$$

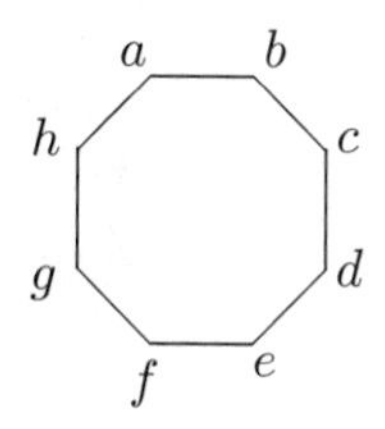

FIGURE 5.1
A Regular Octagon

In D_{16}, the symmetries of a regular 8-gon (an octagon), it works the same, except that there are eight vertices, as illustrated in Figure 5.1. The identity in D_{16} is $e = \begin{pmatrix} a & b & c & d & e & f & g & h \\ a & b & c & d & e & f & g & h \end{pmatrix}$. A one-fourth rotation (two turns) of the octagon results in the symmetry

$$rt = \begin{pmatrix} a & b & c & d & e & f & g & h \\ c & d & e & f & g & h & a & b \end{pmatrix},$$

and reflecting the octagon over a line through the vertices a and e results in the symmetry

$$rf = \begin{pmatrix} a & b & c & d & e & f & g & h \\ a & h & g & f & e & d & c & b \end{pmatrix}.$$

Combining the two symmetries rt and rf using composition of functions results in

$$\begin{aligned} rt \circ rf &= \begin{pmatrix} a & b & c & d & e & f & g & h \\ c & d & e & f & g & h & a & b \end{pmatrix} \circ \begin{pmatrix} a & b & c & d & e & f & g & h \\ a & h & g & f & e & d & c & b \end{pmatrix} \\ &= \begin{pmatrix} a & b & c & d & e & f & g & h \\ g & f & e & d & c & b & a & h \end{pmatrix}, \end{aligned}$$

the symmetry that is a reflection of the octagon over the line passing through the vertices d and h. The inverse of rt is

$$rt^{-1} = \begin{pmatrix} a & b & c & d & e & f & g & h \\ g & h & a & b & c & d & e & f \end{pmatrix},$$

and the inverse of rf is

$$rf^{-1} = \begin{pmatrix} a & b & c & d & e & f & g & h \\ a & h & g & f & e & d & c & b \end{pmatrix} = rf.$$

To conclude our exploration of what constitutes a group, here are a few more examples.

- First, like the previous examples, the set of integers $\mathbb{Z} = \{\ldots, -5, -4, -3, -2, -1, 0, 1, 2, 3, 4, 5, \ldots\}$ with the operation of addition $(+)$ is a group.
- The set of integers $\mathbb{Z} = \{\ldots, -5, -4, -3, -2, -1, 0, 1, 2, 3, 4, 5, \ldots\}$ with the operation of division $(\div)$ is *not* a group since the operation is not closed. For instance, $-3 \div 4 = -\frac{3}{4} = -0.75$, which is not an integer.
- The set of non-negative integers $\mathbb{Z}^* = \{0, 1, 2, 3, 4, 5, \ldots\}$ with the operation of addition $(+)$ is *not* a group. The operation of addition is closed in $\mathbb{Z}^*$, and the

identity element is $e = 0$; however, not every element has an inverse. For instance, the number 5: For 5 to have an inverse, there must be another number a such that $5 + a = 0$. The only solution is for $a = -5$, and -5 is not in $\mathbb{Z}^*$. Thus this is not a group.

Exercises 5.1

1. Perform the indicated operations.

   (a) $3 \oplus_7 6$ (b) $19 \oplus_{25} 22$ (c) $7 \oplus_9 5$ (d) $5 \oplus_{10} 0$

2. Consider the group $\mathbb{Z}_7$ with the operation $\oplus_7$.

   (a) List all the elements in $\mathbb{Z}_7$.

   (b) Give an example to support the fact the operation $\oplus_7$ is associative.

   (c) Find the identity element of $\mathbb{Z}_7$.

   (d) Find the inverse of each non-identity element in $\mathbb{Z}_7$.

3. Consider the group $\mathbb{Z}_{18}$ with the operation $\oplus_{18}$.

   (a) List all the elements in $\mathbb{Z}_{18}$.

   (b) Give an example to support the fact the operation $\oplus_{18}$ is associative.

   (c) Find the identity element of $\mathbb{Z}_{18}$.

   (d) Find the inverse of each non-identity element in $\mathbb{Z}_{18}$.

4. Consider the collection $\mathbb{Z}$ of all integers with the operation of addition (+).

   (a) Find the identity element of $\mathbb{Z}$.

   (b) What is the inverse of 3, of 19, and of -7?

   (c) In general, if n is an integer in $\mathbb{Z}$, what is the inverse of n?

5. Consider S_3, the collection of permutations on the set $\{0, 1, 2\}$, along with the operation of composition of functions ($\circ$) defined on it.

   (a) Find the identity element e of S_3.

   (b) List and clearly label each non-identity element of S_3 and its inverse.

   (c) Given an example to show that the operation defined on S_3 is associative.

6. Consider S_{10}, the collection of permutations on the set $\{0, 1, 2, 3, 4, 5, 6, 7, 8, 9\}$, along with the operation of composition of functions ($\circ$) defined on it.

   (a) Perform the indicated group operations.

   i. $\begin{pmatrix} 0 & 1 & 2 & 3 & 4 & 5 & 6 & 7 & 8 & 9 \\ 3 & 7 & 6 & 1 & 0 & 9 & 2 & 4 & 5 & 8 \end{pmatrix} \circ \begin{pmatrix} 0 & 1 & 2 & 3 & 4 & 5 & 6 & 7 & 8 & 9 \\ 3 & 6 & 9 & 2 & 5 & 8 & 1 & 4 & 7 & 0 \end{pmatrix}$

   ii. $\begin{pmatrix} 0 & 1 & 2 & 3 & 4 & 5 & 6 & 7 & 8 & 9 \\ 2 & 6 & 5 & 0 & 9 & 8 & 1 & 3 & 4 & 7 \end{pmatrix} \circ \begin{pmatrix} 0 & 1 & 2 & 3 & 4 & 5 & 6 & 7 & 8 & 9 \\ 2 & 3 & 4 & 0 & 1 & 9 & 8 & 7 & 6 & 5 \end{pmatrix}$

(b) What is the identity element of S_{10}?

(c) Find the inverse of each element listed below.

i. $\begin{pmatrix} 0 & 1 & 2 & 3 & 4 & 5 & 6 & 7 & 8 & 9 \\ 3 & 7 & 6 & 1 & 0 & 9 & 2 & 4 & 5 & 8 \end{pmatrix}$

ii. $\begin{pmatrix} 0 & 1 & 2 & 3 & 4 & 5 & 6 & 7 & 8 & 9 \\ 2 & 6 & 5 & 0 & 9 & 8 & 1 & 3 & 4 & 7 \end{pmatrix}$

7. Consider the dihedral group D_8, the collection of symmetries of a regular 4-gon (a square).

   (a) List all eight elements of D_8.

   (b) Find the identity element e of D_8.

   (c) Find the inverse of each of the non-identity elements in D_8.

   (d) Give an example to show that the operation defined on D_8 is associative.

8. Consider the dihedral group D_{12}, the collection of symmetries of a regular 6-gon (a hexagon).

   (a) List all 12 elements of D_{12}.

   (b) Find the identity element e of D_{12}.

   (c) Find the inverse of each of the non-identity elements in D_{12}.

9. Show that the following sets with the operations indicated are *not* groups. Provide specific calculations to support your work.

   (a) $\mathbb{Z}_{10}$ with the operation $\oplus_{15}$.

   (b) The set $\{0, 2, 3, 4, 6, 8\}$ with the operation $\oplus_{10}$.

   (c) The set of negative integers $\mathbb{Z}^- = \{\ldots, -5, -4, -3, -2, -1\}$ with the operation of subtraction $(-)$.

   (d) The set of positive integers $\mathbb{Z}^+ = \{1, 2, 3, 4, 5, \ldots\}$ with the operation of addition $(+)$.

   (e) The set of integers without 0, or $\{\ldots, -4, -3, -2, -1, 1, 2, 3, 4, \ldots\}$ with the operation of addition $(+)$.

10. Show that each of the following sets with the indicated operation *is* a group. You must show how each of the four criteria from Definition 5.1.8 is met so that the set and the indicated operation form a group.

    (a) $\{0, 2, 4, 6, 8\}$ with the operation $\oplus_{10}$.

    (b) $\{0, 3, 6, 9, 12\}$ with the operation $\oplus_{15}$.

    (c) The collection of symmetries on a rectangle (as drawn below).

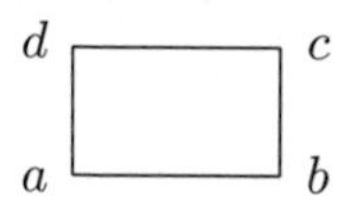

(d) The collection of rational numbers $\mathbb{Q}$ with the operation of addition $(+)$.

(e) The collection of non-zero rational numbers from $\mathbb{Q}$ (all rational numbers except 0) with the operation of multiplication $(\cdot)$.

Paper Assignments 5.1

1. **Analysis and Serializing.** Definition 5.1.8 presents the four conditions that a set with an operation defined on that set must satisfy in order for the set to be called a *group*. Many examples of groups were presented and discussed.

   Another example is the set $2\mathbb{Z} = \{\ldots, -8, -6, -4, -2, 0, 2, 4, 6, 8, \ldots\}$ (the set of even integers), with the operation of addition $(+)$. In the order listed in Definition 5.1.8, show how the set $2\mathbb{Z}$, along with the operation of addition, satisfies each of the conditions needed for it to be a group. In other words, demonstrate that $2\mathbb{Z}$ is a group by showing that it satisfies each of the conditions in Definition 5.1.8.

2. **Analysis and Serializing.** Definition 5.1.8 presents the four conditions that a set with an operation defined on that set must satisfy in order for the set to be called a *group*. Many examples of groups were presented and discussed.

   Another example is the set of symmetries of a rectangle (shown below), along with the operation of composition.

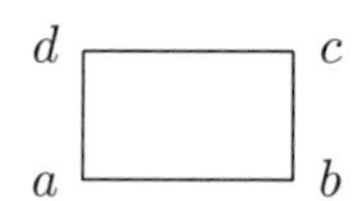

   (a) Find all the symmetries of the rectangle presented above.

   (b) In the same order as listed in Definition 5.1.8, show how the set of symmetries found in (a), along with the operation of composition, satisfies each of the conditions needed for it to be a group. In other words, demonstrate that the set of symmetries forms a group by showing that it satisfies each of the conditions in Definition 5.1.8.

Group Activity 5.1

1. The set $\mathbb{Z}_n = \{0, 1, 2, \ldots, n-1\}$, along with the operation $\oplus_n$, is a group.

   (a) Each member of the group chooses one of the following groups and lists each element and its inverse: $\mathbb{Z}_{20}$, $\mathbb{Z}_{21}$, $\mathbb{Z}_{25}$, $\mathbb{Z}_{30}$.

   (b) Comparing your work in part (a), do you notice a relation between each element of the group and its inverse? Does there seem to be a formula or method that could be followed to obtain each element's inverse?

   (c) Write a formula, one for each of the groups $\mathbb{Z}_{20}$, $\mathbb{Z}_{21}$, $\mathbb{Z}_{25}$, and $\mathbb{Z}_{30}$, that can be used to quickly find the inverse of each element from that particular group.

(d) Generalize your formulas from part(c) to one that works for any group $\mathbb{Z}_n$ with the operation $\oplus_n$. Use this formula to find the inverses of each element in $\mathbb{Z}_{50}$.

Further Reading

Hannabuss, K., Sound and Symmetry, *Mathematical Intelligencer*, 19(4), 1997, 16–21.

Humphreys, J. F., *A Course in Group Theory,* Oxford University Press, New York City, 1997.

Isihara, P., and Knapp, M., Basic Z_{12} Analysis of Musical Chords, *UMAP Journal*, 14(4), 1993, 319–348.

White, A. T., Fabian Stedman: The First Group Theorist?, *American Mathematical Monthly*, 103(9), 1996, 771–778.

5.2 Cayley Tables

Preliminary Activity. Consider the group $\mathbb{Z}_8 = \{0, 1, 2, 3, 4, 5, 6, 7\}$ with the operation of $\oplus_8$. Create a table, using any method you deem appropriate, that will list all of the elements of $\mathbb{Z}_8$ and the result of combining any pair of elements from $\mathbb{Z}_8$ using $\oplus_8$.

Given a finite group G with an operation $*$ on G, there is a way to visualize all the different elements of G and all the different ways to combine them using the operation $*$. All the different elements of G and the different ways to combine them can be written in a single table, called a *Cayley table*, for easy reference.

To create a Cayley table for a group G with operation $*$, first list all the elements of G along the top and again down the left side of the table, being sure to list them in the same order in both places. In any list, the identity element e usually goes first. For two elements a and b in G, their combination $a * b$ is listed in the ath row and bth column of the table, as shown in Table 5.1.

TABLE 5.1
General Cayley Table

$*$	$\cdots$	$\cdots$	b	$\cdots$	$\cdots$
$\vdots$			$\vdots$		
$\vdots$			$\vdots$		
a	$\cdots$	$\cdots$	$a * b$		
$\vdots$					
$\vdots$					

TABLE 5.2
Cayley Table for $\mathbb{Z}_4$

$\oplus_4$	0	1	2	3
0	0	1	2	3
1	1	2	3	0
2	2	3	0	1
3	3	0	1	2

For example, Table 5.2 shows the Cayley table for the finite group $\mathbb{Z}_4$ with operation $\oplus_4$. The elements 0, 1, 2, and 3 are listed, in the same order, across the top and down the left side of the table. The result of combining each pair of elements from $\mathbb{Z}_4$ is then listed in the body of the table. For instance, the combination $2 \oplus_4 1 = 3$, thus 3 goes in the row starting with 2 and the column starting with 1, as indicated on the left in Table 5.3. Similarly, the result of $2 \oplus_4 3 = 1$ goes in the row starting with 2 and the column starting with 3, as indicated on the right.

TABLE 5.3
Cayley Table for $\mathbb{Z}_4$

			↓		
	$\oplus_4$	0	**1**	2	3
	0	0	1	2	3
	1	1	2	3	0
→	**2**	2	**3**	0	1
	3	3	0	1	2

					↓
	$\oplus_4$	0	1	2	**3**
	0	0	1	2	3
	1	1	2	3	0
→	**2**	2	3	0	**1**
	3	3	0	1	2

Test yourself by creating the Cayley table for group $\mathbb{Z}_5$. The result should be the Cayley table presented in Table 5.4.

TABLE 5.4
Cayley Table for $\mathbb{Z}_5$

$\oplus_4$	0	1	2	3	4
0	0	1	2	3	4
1	1	2	3	4	0
2	2	3	4	0	1
3	3	4	0	1	2
4	4	0	1	2	3

The Cayley table for any finite group can be created. Just follow the same procedure and be sure to combine the group elements using the operation of the group.

Example 5.2.1. *Consider the group S_3 with the composition operation $\circ$. Construct its Cayley table.*

- To make it easier to list the group elements in the Cayley table, represent the elements of S_3 in the following manner:

$$0 = \begin{pmatrix} 0 & 1 & 2 \\ 0 & 1 & 2 \end{pmatrix} \qquad 1 = \begin{pmatrix} 0 & 1 & 2 \\ 1 & 2 & 0 \end{pmatrix} \qquad 2 = \begin{pmatrix} 0 & 1 & 2 \\ 2 & 0 & 1 \end{pmatrix}$$

$$3 = \begin{pmatrix} 0 & 1 & 2 \\ 0 & 2 & 1 \end{pmatrix} \qquad 4 = \begin{pmatrix} 0 & 1 & 2 \\ 2 & 1 & 0 \end{pmatrix} \qquad 5 = \begin{pmatrix} 0 & 1 & 2 \\ 1 & 0 & 2 \end{pmatrix}$$

- Given these representations of the elements in S_3, Table 5.5 presents the Cayley table for S_3.

TABLE 5.5
Cayley Table for S_3

$\circ$	0	1	2	3	4	5
0	0	1	2	3	4	5
1	1	2	0	4	5	3
2	2	0	1	5	3	4
3	3	5	4	0	2	1
4	4	3	5	1	0	2
5	5	4	3	2	1	0

- For example,

$$\begin{aligned} 4 \circ 3 &= \begin{pmatrix} 0 & 1 & 2 \\ 2 & 1 & 0 \end{pmatrix} \circ \begin{pmatrix} 0 & 1 & 2 \\ 0 & 2 & 1 \end{pmatrix} \\ &= \begin{pmatrix} 0 & 1 & 2 \\ 1 & 2 & 0 \end{pmatrix} \\ &= 1. \end{aligned}$$

Example 5.2.2. *Create the Cayley table for the dihedral group D_8, the collection of symmetries of a regular 4-gon (a square), using the elements indicated below.*

d c

a b

- As with S_3, to make the creation of the table easier, use the following representations of the elements of D_8.

$$0 = \begin{pmatrix} a & b & c & d \\ a & b & c & d \end{pmatrix} \quad 1 = \begin{pmatrix} a & b & c & d \\ b & c & d & a \end{pmatrix} \quad 2 = \begin{pmatrix} a & b & c & d \\ c & d & a & b \end{pmatrix}$$

$$3 = \begin{pmatrix} a & b & c & d \\ d & a & b & c \end{pmatrix} \quad 4 = \begin{pmatrix} a & b & c & d \\ b & a & d & c \end{pmatrix} \quad 5 = \begin{pmatrix} a & b & c & d \\ d & c & b & a \end{pmatrix}$$

$$6 = \begin{pmatrix} a & b & c & d \\ a & d & c & b \end{pmatrix} \quad 7 = \begin{pmatrix} a & b & c & d \\ c & b & a & d \end{pmatrix}$$

- Table 5.6 gives the Cayley table for D_8.

TABLE 5.6
Cayley Table for D_8

$\circ$	0	1	2	3	4	5	6	7
0	0	1	2	3	4	5	6	7
1	1	2	3	0	6	7	5	4
2	2	3	0	1	5	4	7	6
3	3	0	1	2	7	6	4	5
4	4	7	5	6	0	2	3	1
5	5	6	4	7	2	0	1	3
6	6	4	7	5	1	3	0	2
7	7	5	6	4	3	1	2	0

- For example,

$$\begin{aligned} 3 \circ 6 &= \begin{pmatrix} a & b & c & d \\ d & a & b & c \end{pmatrix} \circ \begin{pmatrix} a & b & c & d \\ a & d & c & b \end{pmatrix} \\ &= \begin{pmatrix} a & b & c & d \\ b & a & d & c \end{pmatrix} \\ &= 4. \end{aligned}$$

As previously mentioned, the Cayley table for a group provides an easy way to read the elements of the group and to see how to combine them using the operation of the group. If a Cayley table presents a group, then the identity element can be easily identified, as can the inverses for each of the elements of the group. Suppose that the set $H = \{\alpha, \beta, \gamma, \delta\}$ with an operation $*$ form a group. The Cayley table for H, presented in Table 5.7, gives all the needed information concerning the group. Don't be confused by the group elements α, β, γ, and δ. They are just the symbols for the four elements of this group. From the table, it can be determined that $\alpha * \gamma = \gamma$, that $\beta * \gamma = \delta$, and that $\gamma * \gamma = \alpha$.

TABLE 5.7
Cayley Table for H

$*$	α	β	γ	δ
α	α	β	γ	δ
β	β	γ	δ	α
γ	γ	δ	α	β
δ	δ	α	β	γ

To find the identity element of H, we must find the element e of H such that $e * a = a * e = a$ for each element $a \in H$. There are only four possibilities for e: e must be α, δ, γ, or δ. By using the Cayley table, it is readily seen that the only element that satisfies the stated condition is α. Thus the identity element $e = \alpha$.

Now that it is known that the identity element of H is α, the inverse of each element in H can be found. For each element $a \in H$, a^{-1} will satisfy $a * a^{-1} = a^{-1} * a = \alpha$ (the identity element of H). Like the identity element, the inverse can be obtained by looking at the Cayley table.

Let's start with α itself. It is an element of H and must have an inverse α^{-1} such that $\alpha * \alpha^{-1} = \alpha$. Locating α on the left-hand column of the Cayley table, find the element listed on the top of the table, such that $\alpha * \alpha^{-1} = \alpha$. The only possible answer is that $\alpha^{-1} = \alpha$, the element below the check-mark $\surd$ in Table 5.8 on the left. A quick check shows that $\alpha^{-1} * \alpha = \alpha * \alpha$ also equals the identity α.

Similiarly, to find the inverse of β, the element β^{-1} must be found that satisfies the equation $\beta * \beta^{-1} = \beta^{-1} * \beta = \alpha$ (recall that α is the identity element of H). Locating β on the left-hand column of the Cayley table, find the element listed on the top of the table, such that $\beta * \beta^{-1} = \alpha$. The only possible answer is that $\beta^{-1} = \delta$, the element below the check-mark $\surd$ in Table 5.8 on the right. It is easy to check that $\beta^{-1} * \beta = \delta * \beta = \alpha$.

Finding γ^{-1} and δ^{-1} works the same way. It is simple to show that $\gamma^{-1} = \gamma$ and $\delta^{-1} = \beta$.

TABLE 5.8
Locating Identities and Inverses in the Cayley Table for H

		$\surd$			
	$*$	$\boldsymbol{\alpha}$	β	γ	δ
$\rightarrow$	$\boldsymbol{\alpha}$	$\boldsymbol{\alpha}$	β	γ	δ
	β	β	γ	δ	α
	γ	γ	δ	α	β
	δ	δ	α	β	γ

					$\surd$
	$*$	α	β	γ	$\boldsymbol{\delta}$
	α	α	β	γ	δ
$\rightarrow$	$\boldsymbol{\beta}$	β	γ	δ	$\boldsymbol{\alpha}$
	γ	γ	δ	α	β
	δ	δ	α	β	γ

Exercises 5.2

1. The collection of elements $G = \{\bigstar, \sharp, \P, \S, \dagger, \exists\}$ and the binary operation $*$, defined in the accompanying Cayley table, form a group.

$*$	$\bigstar$	$\sharp$	$\P$	$\S$	$\dagger$	$\exists$
$\bigstar$	$\bigstar$	$\sharp$	$\P$	$\S$	$\dagger$	$\exists$
$\sharp$	$\sharp$	$\P$	$\bigstar$	$\dagger$	$\exists$	$\S$
$\P$	$\P$	$\bigstar$	$\sharp$	$\exists$	$\S$	$\dagger$
$\S$	$\S$	$\exists$	$\dagger$	$\bigstar$	$\P$	$\sharp$
$\dagger$	$\dagger$	$\S$	$\exists$	$\sharp$	$\bigstar$	$\P$
$\exists$	$\exists$	$\dagger$	$\S$	$\P$	$\sharp$	$\bigstar$

(a) Based on the Cayley table for G, what would be the results of these calculations?

(b) Find the identity element for this group.

(c) Find the inverse of each element in this group.

2. Create the Cayley table for the group $\mathbb{Z}_7$.

3. Consider the collection of symmetries on a rectangle, as shown. You showed in Section 5.1 that this collection is a group. Create the Cayley table for this group.

d c

a b

4. Create the Cayley table for $\mathbb{Z}_{10}$.

5. Create the Cayley table for the dihedral group D_{10}, the collection of symmetries on a regular 5-gon (a pentagon). To make the creation of the table easier, use the following representations of the elements of D_{10}.

$$0 = \begin{pmatrix} a & b & c & d & e \\ a & b & c & d & e \end{pmatrix} \quad 1 = \begin{pmatrix} a & b & c & d & e \\ b & c & d & e & a \end{pmatrix} \quad 2 = \begin{pmatrix} a & b & c & d & e \\ c & d & e & a & b \end{pmatrix}$$

$$3 = \begin{pmatrix} a & b & c & d & e \\ d & e & a & b & c \end{pmatrix} \quad 4 = \begin{pmatrix} a & b & c & d & e \\ e & a & b & c & d \end{pmatrix} \quad 5 = \begin{pmatrix} a & b & c & d & e \\ a & e & d & c & b \end{pmatrix}$$

$$6 = \begin{pmatrix} a & b & c & d & e \\ e & d & c & b & a \end{pmatrix} \quad 7 = \begin{pmatrix} a & b & c & d & e \\ d & c & b & a & e \end{pmatrix} \quad 8 = \begin{pmatrix} a & b & c & d & e \\ c & b & a & e & d \end{pmatrix}$$

$$9 = \begin{pmatrix} a & b & c & d & e \\ b & a & e & d & c \end{pmatrix}$$

Paper Assignment 5.2

1. **Analysis.** Write a paper that presents the work done in Group Activity 5.2.1.

Group Activities 5.2

1. A group G with operation $*$ is *abelian* if $a * b = b * a$ for all elements a and b in G. In other words, G is *abelian* if the operation defined on G is commutative. For example, the group $\mathbb{Z} = \{\ldots, -4, -3, -2, -1, 0, 1, 2, 3, 4, \ldots\}$, with the operation of addition (+) is abelian ($a + b = b + a$ for all integers a and b), while the dihedral group D_8, presented in Table 5.6, is not ($5 = 1 \circ 6 \neq 6 \circ 1 = 4$).

   In Section 5.1, it was shown that both $\mathbb{Z}_n = \{0, 1, 2, \ldots, n-1\}$, with the operation $\oplus_n$, and S_n, the collection of permutations of the set $\{0, 1, 2, \ldots, n-1\}$ with the operation composition of function $\circ$, are both groups. Show that $\mathbb{Z}_n$ is an abelian group for all values of n and that S_n is not abelian for $n \geq 3$. To do this, first show that $\mathbb{Z}_3$ and $\mathbb{Z}_4$ are abelian. Then generalize your work to show that $\mathbb{Z}_n$ is abelian. After that, show that S_3 and S_4 are not abelian. Then generalize your work to show that S_n is not abelian.

2. (a) Show that this table cannot be the Cayley table for the group $G = \{a, b, c\}$.

$*$	a	b	c
a	a	b	c
b	b	a	a
c	c	b	b

   (b) Show that this table cannot be the Cayley table for the group $G = \{a, b, c\}$.

$*$	a	b	c
a	a	c	b
b	c	b	a
c	b	a	c

   (c) In parts (a) and (b), each table was not a Cayley table for group $G = \{a, b, c\}$, because a fundamental group property was not satisfied. Use the properties of a group (recall Definition 5.1.8) to complete the following Cayley table for the group $G = \{a, b, c, d\}$.

$*$	a	b	c	d
a	a	b	c	
b		c	d	a
c	c	d		b
d	d	a	b	

Further Reading

Hannabuss, K., Sound and Symmetry, *Mathematical Intelligencer*, 19(4), 1997, 16–21.

Humphreys, J. F., *A Course in Group Theory,* Oxford University Press, New York City, 1997.

White, A. T., Fabian Stedman: The First Group Theorist?, *American Mathematical Monthly*, 103(9), 1996, 771–778.

5.3 Powers and Orders of Group Elements

The work done in Sections 5.1 and 5.2 on groups and Cayley tables will now be used to investigate powers and orders of group elements. Once this is complete, the last piece will be in place so that the new check digit scheme promised at the beginning of this chapter can be developed.

Preliminary Activity. Consider the following elements from S_5.

$$\sigma_1 = \begin{pmatrix} 0 & 1 & 2 & 3 & 4 \\ 1 & 2 & 3 & 4 & 0 \end{pmatrix} \quad \sigma_2 = \begin{pmatrix} 0 & 1 & 2 & 3 & 4 \\ 2 & 1 & 4 & 3 & 0 \end{pmatrix} \quad \sigma_3 = \begin{pmatrix} 0 & 1 & 2 & 3 & 4 \\ 1 & 0 & 3 & 2 & 4 \end{pmatrix}$$

1. Compute.
   (a) $\Big(\big((\sigma_1 \circ \sigma_1) \circ \sigma_1\big) \circ \sigma_1\Big) \circ \sigma_1$
   (b) $(\sigma_2 \circ \sigma_2) \circ \sigma_2$
   (c) $\sigma_3 \circ \sigma_3$
2. Develop or find a shorthand way to represent each of the computations from part (1).
3. What is the result of each computation from part (1)? Is it a special element from S_5?

Consider the two elements

$$\sigma = \begin{pmatrix} 0 & 1 & 2 & 3 & 4 & 5 \\ 2 & 3 & 0 & 5 & 4 & 1 \end{pmatrix} \quad \text{and} \quad \tau = \begin{pmatrix} 0 & 1 & 2 & 3 & 4 & 5 \\ 3 & 4 & 5 & 0 & 1 & 2 \end{pmatrix})$$

from the symmetric group S_6 and each of the following computations:

$$\begin{aligned} \sigma \circ \tau &= \begin{pmatrix} 0 & 1 & 2 & 3 & 4 & 5 \\ 2 & 3 & 0 & 5 & 4 & 1 \end{pmatrix} \circ \begin{pmatrix} 0 & 1 & 2 & 3 & 4 & 5 \\ 3 & 4 & 5 & 0 & 1 & 2 \end{pmatrix} \\ &= \begin{pmatrix} 0 & 1 & 2 & 3 & 4 & 5 \\ 5 & 0 & 3 & 2 & 1 & 4 \end{pmatrix}, \end{aligned}$$

$$\begin{aligned}\sigma \circ \sigma &= \begin{pmatrix} 0 & 1 & 2 & 3 & 4 & 5 \\ 2 & 3 & 0 & 5 & 4 & 1 \end{pmatrix} \circ \begin{pmatrix} 0 & 1 & 2 & 3 & 4 & 5 \\ 2 & 3 & 0 & 5 & 4 & 1 \end{pmatrix} \\ &= \begin{pmatrix} 0 & 1 & 2 & 3 & 4 & 5 \\ 0 & 5 & 2 & 1 & 4 & 3 \end{pmatrix},\end{aligned}$$

$$\begin{aligned}\tau \circ \tau &= \begin{pmatrix} 0 & 1 & 2 & 3 & 4 & 5 \\ 3 & 4 & 5 & 0 & 1 & 2 \end{pmatrix} \circ \begin{pmatrix} 0 & 1 & 2 & 3 & 4 & 5 \\ 3 & 4 & 5 & 0 & 1 & 2 \end{pmatrix} \\ &= \begin{pmatrix} 0 & 1 & 2 & 3 & 4 & 5 \\ 0 & 1 & 2 & 3 & 4 & 5 \end{pmatrix}.\end{aligned}$$

The second computation $\sigma \circ \sigma$ can be expressed as σ^2 and the third as $\tau^2 = \tau \circ \tau$. The positive integer 2 in the exponent indicates that the element is to be combined with itself twice using the operation $\circ$ (composition of functions). However, an element can be raised to any power.

Definition 5.3.1. *Let g be an element of a group G with operation $*$, and let n be a positive integer. Then*

$$g^n = \underbrace{g * g * \cdots * g}_{n \text{ times}}.$$

An element raised to the first power equals itself; g^1 always equals g. Because the operation $*$ of any group is associative, $*$ can be applied to combine the elements in any order. In groups where addition or $\oplus_n$ is the group operation, the notation $n \cdot g$ is used instead of g^n. In this case,

$$n \cdot g = \underbrace{g + g + \cdots + g}_{n \text{ times}}.$$

Example 5.3.2. *Consider*

$$f = \begin{pmatrix} 0 & 1 & 2 & 3 & 4 \\ 3 & 4 & 0 & 1 & 2 \end{pmatrix} \in S_5.$$

$$\begin{array}{lccccc} f^3 = & f & \circ & f & \circ & f \\ = & \begin{pmatrix} 0 & 1 & 2 & 3 & 4 \\ 3 & 4 & 0 & 1 & 2 \end{pmatrix} & \circ & \begin{pmatrix} 0 & 1 & 2 & 3 & 4 \\ 3 & 4 & 0 & 1 & 2 \end{pmatrix} & \circ & \begin{pmatrix} 0 & 1 & 2 & 3 & 4 \\ 3 & 4 & 0 & 1 & 2 \end{pmatrix} \\ = & \left(\begin{pmatrix} 0 & 1 & 2 & 3 & 4 \\ 3 & 4 & 0 & 1 & 2 \end{pmatrix}\right. & \circ & \left.\begin{pmatrix} 0 & 1 & 2 & 3 & 4 \\ 3 & 4 & 0 & 1 & 2 \end{pmatrix}\right) & \circ & \begin{pmatrix} 0 & 1 & 2 & 3 & 4 \\ 3 & 4 & 0 & 1 & 2 \end{pmatrix} \\ = & \searrow & \downarrow & \swarrow & \downarrow & \downarrow \\ = & & \begin{pmatrix} 0 & 1 & 2 & 3 & 4 \\ 1 & 2 & 3 & 4 & 0 \end{pmatrix} & & \circ & \begin{pmatrix} 0 & 1 & 2 & 3 & 4 \\ 3 & 4 & 0 & 1 & 2 \end{pmatrix} \\ = & & & \begin{pmatrix} 0 & 1 & 2 & 3 & 4 \\ 4 & 0 & 1 & 2 & 3 \end{pmatrix}. & & \end{array}$$

Example 5.3.3. *Consider*

$$\sigma = \begin{pmatrix} 0 & 1 & 2 & 3 & 4 & 5 \\ 1 & 2 & 3 & 4 & 0 & 5 \end{pmatrix} \in S_6.$$

$$\begin{aligned}
\sigma^4 &= \sigma \circ \sigma \circ \sigma \circ \sigma \\
&= \begin{pmatrix} 0 & 1 & 2 & 3 & 4 & 5 \\ 1 & 2 & 3 & 4 & 0 & 5 \end{pmatrix} \circ \begin{pmatrix} 0 & 1 & 2 & 3 & 4 & 5 \\ 1 & 2 & 3 & 4 & 0 & 5 \end{pmatrix} \circ \begin{pmatrix} 0 & 1 & 2 & 3 & 4 & 5 \\ 1 & 2 & 3 & 4 & 0 & 5 \end{pmatrix} \\
&\qquad \circ \begin{pmatrix} 0 & 1 & 2 & 3 & 4 & 5 \\ 1 & 2 & 3 & 4 & 0 & 5 \end{pmatrix} \\
&= \left(\begin{pmatrix} 0 & 1 & 2 & 3 & 4 & 5 \\ 1 & 2 & 3 & 4 & 0 & 5 \end{pmatrix} \circ \begin{pmatrix} 0 & 1 & 2 & 3 & 4 & 5 \\ 1 & 2 & 3 & 4 & 0 & 5 \end{pmatrix} \right) \\
&\qquad \circ \left(\begin{pmatrix} 0 & 1 & 2 & 3 & 4 & 5 \\ 1 & 2 & 3 & 4 & 0 & 5 \end{pmatrix} \circ \begin{pmatrix} 0 & 1 & 2 & 3 & 4 & 5 \\ 1 & 2 & 3 & 4 & 0 & 5 \end{pmatrix} \right) \\
&= \begin{pmatrix} 0 & 1 & 2 & 3 & 4 & 5 \\ 2 & 3 & 4 & 0 & 1 & 5 \end{pmatrix} \circ \begin{pmatrix} 0 & 1 & 2 & 3 & 4 & 5 \\ 2 & 3 & 4 & 0 & 1 & 5 \end{pmatrix} \\
&= \begin{pmatrix} 0 & 1 & 2 & 3 & 4 & 5 \\ 4 & 0 & 1 & 2 & 3 & 5 \end{pmatrix}.
\end{aligned}$$

When performing these types of computations, a permutation may be given as a product of cycles. However, for raising an element expressed as a product of cycles to a particular power, it is often easier to write the results of each calculation in permutation notation. The answer can be written in either form.

Example 5.3.4. *Consider* $\tau = (0,2,4)(1,6,5,3) \in S_7$.

$$\begin{aligned}
\tau^5 &= \tau \circ \tau \circ \tau \circ \tau \circ \tau \\
&= (0,2,4)(1,6,5,3) \circ (0,2,4)(1,6,5,3) \circ (0,2,4)(1,6,5,3) \circ (0,2,4)(1,6,5,3) \\
&\qquad \circ (0,2,4)(1,6,5,3) \\
&= ((0,2,4)(1,6,5,3) \circ (0,2,4)(1,6,5,3)) \\
&\qquad \circ ((0,2,4)(1,6,5,3) \circ (0,2,4)(1,6,5,3)) \circ (0,2,4)(1,6,5,3) \\
&= \begin{pmatrix} 0 & 1 & 2 & 3 & 4 & 5 & 6 \\ 4 & 5 & 0 & 6 & 2 & 1 & 3 \end{pmatrix} \circ \begin{pmatrix} 0 & 1 & 2 & 3 & 4 & 5 & 6 \\ 4 & 5 & 0 & 6 & 2 & 1 & 3 \end{pmatrix} \circ (0,2,4)(1,6,5,3) \\
&= \left(\begin{pmatrix} 0 & 1 & 2 & 3 & 4 & 5 & 6 \\ 4 & 5 & 0 & 6 & 2 & 1 & 3 \end{pmatrix} \circ \begin{pmatrix} 0 & 1 & 2 & 3 & 4 & 5 & 6 \\ 4 & 5 & 0 & 6 & 2 & 1 & 3 \end{pmatrix} \right) \circ (0,2,4)(1,6,5,3) \\
&= \begin{pmatrix} 0 & 1 & 2 & 3 & 4 & 5 & 6 \\ 2 & 1 & 4 & 3 & 0 & 5 & 6 \end{pmatrix} \circ (0,2,4)(1,6,5,3) \\
&= \begin{pmatrix} 0 & 1 & 2 & 3 & 4 & 5 & 6 \\ 4 & 6 & 0 & 1 & 2 & 3 & 5 \end{pmatrix} \\
&= (0,4,2)(1,6,5,3).
\end{aligned}$$

Example 5.3.5. *Consider the group $\mathbb{Z}$ with the operation of addition, $(+)$.*

- $3 \cdot 7 = 7 + 7 + 7 = 21.$
- $5 \cdot (-2) = (-2) + (-2) + (-2) + (-2) + (-2) = -10.$

Example 5.3.6. *Let g be an element of $\mathbb{Z}_n$ with operation $\oplus_n$. Given the work on modulo arithmetic done in Chapter 2, for any positive integer m,*

$$\begin{aligned} m \cdot g &= g \oplus_n g \oplus_n g \oplus_n \cdots \oplus_n g \\ &= (g + g + g + \cdots + g) \pmod{n}. \end{aligned}$$

- For the element $8 \in \mathbb{Z}_{10}$ with the operation $\oplus_{10}$,

$$\begin{aligned} 3 \cdot 8 &= 8 \oplus_{10} 8 \oplus_{10} 8 \\ &= (8 + 8 + 8) \pmod{10} \\ &= 24 \pmod{10} \\ &= 4. \end{aligned}$$

- For the element $5 \in \mathbb{Z}_{12}$ with the operation $\oplus_{12}$,

$$\begin{aligned} 4 \cdot 5 &= 5 \oplus_{12} 5 \oplus_{12} 5 \oplus_{12} 5 \\ &= (5 + 5 + 5 + 5) \pmod{12} \\ &= 20 \pmod{12} \\ &= 8. \end{aligned}$$

Recall the element $\tau = \left(\begin{smallmatrix} 0 & 1 & 2 & 3 & 4 & 5 \\ 3 & 4 & 5 & 0 & 1 & 2 \end{smallmatrix}\right)$ from the symmetric group S_6 presented at the beginning of this section. The computation

$$\begin{aligned} \tau^2 &= \tau \circ \tau \\ &= \begin{pmatrix} 0 & 1 & 2 & 3 & 4 & 5 \\ 3 & 4 & 5 & 0 & 1 & 2 \end{pmatrix} \circ \begin{pmatrix} 0 & 1 & 2 & 3 & 4 & 5 \\ 3 & 4 & 5 & 0 & 1 & 2 \end{pmatrix} \\ &= \begin{pmatrix} 0 & 1 & 2 & 3 & 4 & 5 \\ 0 & 1 & 2 & 3 & 4 & 5 \end{pmatrix} \\ &= e, \end{aligned}$$

the identity element of S_6.

Definition 5.3.7. *Let G be a group and g be an element in G. The* **order** *of $g \in G$, denoted $\mid g \mid$, is the smallest positive integer n, such that $g^n = e$, the identity element of G. If no such n exists, then g has infinite order.*

For the permutation $\tau \in S_6$, the order of τ or $\mid \tau \mid = 2$. Now, $\tau^4 = e$ as well. However $\mid \tau \mid = 2$, since 2 is the smallest positive integer n such that $\tau^n = e$. In groups where addition or $\oplus_n$ is used as the group operation, the order of $g \in G$ is the smallest positive integer such that $n \cdot g = e$, the identity element of G. The order of a finite group G, denoted $\mid G \mid$, is the number of elements in G.

Example 5.3.8. *For* $\sigma = \left(\begin{smallmatrix} 0 & 1 & 2 & 3 & 4 & 5 \\ 1 & 2 & 3 & 4 & 0 & 5 \end{smallmatrix}\right) \in S_6$, $\mid \sigma \mid = 5$ *since* 5 *is the smallest positive integer* n *such that* $\sigma^n = e = \left(\begin{smallmatrix} 0 & 1 & 2 & 3 & 4 & 5 \\ 0 & 1 & 2 & 3 & 4 & 5 \end{smallmatrix}\right)$. *Simple calculations show that* $\sigma^2 \neq e$, $\sigma^3 \neq e$, *and* $\sigma^4 \neq e$. *However:*

$$
\begin{aligned}
\sigma^5 &= \sigma \circ \sigma \circ \sigma \circ \sigma \circ \sigma \\
&= \begin{pmatrix} 0 & 1 & 2 & 3 & 4 & 5 \\ 1 & 2 & 3 & 4 & 0 & 5 \end{pmatrix} \circ \begin{pmatrix} 0 & 1 & 2 & 3 & 4 & 5 \\ 1 & 2 & 3 & 4 & 0 & 5 \end{pmatrix} \circ \begin{pmatrix} 0 & 1 & 2 & 3 & 4 & 5 \\ 1 & 2 & 3 & 4 & 0 & 5 \end{pmatrix} \\
&\qquad \circ \begin{pmatrix} 0 & 1 & 2 & 3 & 4 & 5 \\ 1 & 2 & 3 & 4 & 0 & 5 \end{pmatrix} \circ \begin{pmatrix} 0 & 1 & 2 & 3 & 4 & 5 \\ 1 & 2 & 3 & 4 & 0 & 5 \end{pmatrix} \\
&= \left(\begin{pmatrix} 0 & 1 & 2 & 3 & 4 & 5 \\ 1 & 2 & 3 & 4 & 0 & 5 \end{pmatrix} \circ \begin{pmatrix} 0 & 1 & 2 & 3 & 4 & 5 \\ 1 & 2 & 3 & 4 & 0 & 5 \end{pmatrix} \right) \\
&\qquad \circ \left(\begin{pmatrix} 0 & 1 & 2 & 3 & 4 & 5 \\ 1 & 2 & 3 & 4 & 0 & 5 \end{pmatrix} \circ \begin{pmatrix} 0 & 1 & 2 & 3 & 4 & 5 \\ 1 & 2 & 3 & 4 & 0 & 5 \end{pmatrix} \right) \\
&\qquad \circ \begin{pmatrix} 0 & 1 & 2 & 3 & 4 & 5 \\ 1 & 2 & 3 & 4 & 0 & 5 \end{pmatrix} \\
&= \begin{pmatrix} 0 & 1 & 2 & 3 & 4 & 5 \\ 2 & 3 & 4 & 0 & 1 & 5 \end{pmatrix} \circ \begin{pmatrix} 0 & 1 & 2 & 3 & 4 & 5 \\ 2 & 3 & 4 & 0 & 1 & 5 \end{pmatrix} \circ \begin{pmatrix} 0 & 1 & 2 & 3 & 4 & 5 \\ 1 & 2 & 3 & 4 & 0 & 5 \end{pmatrix} \\
&= \left(\begin{pmatrix} 0 & 1 & 2 & 3 & 4 & 5 \\ 2 & 3 & 4 & 0 & 1 & 5 \end{pmatrix} \circ \begin{pmatrix} 0 & 1 & 2 & 3 & 4 & 5 \\ 2 & 3 & 4 & 0 & 1 & 5 \end{pmatrix} \right) \circ \begin{pmatrix} 0 & 1 & 2 & 3 & 4 & 5 \\ 1 & 2 & 3 & 4 & 0 & 5 \end{pmatrix} \\
&= \begin{pmatrix} 0 & 1 & 2 & 3 & 4 & 5 \\ 4 & 0 & 1 & 2 & 3 & 5 \end{pmatrix} \circ \begin{pmatrix} 0 & 1 & 2 & 3 & 4 & 5 \\ 1 & 2 & 3 & 4 & 0 & 5 \end{pmatrix} \\
&= \begin{pmatrix} 0 & 1 & 2 & 3 & 4 & 5 \\ 0 & 1 & 2 & 3 & 4 & 5 \end{pmatrix} \\
&= e.
\end{aligned}
$$

Example 5.3.9. *The order of the element* $3 \in \mathbb{Z}_7$ *is* 7, *or* $\mid 3 \mid = 7$, *as* 7 *is the smallest positive integer* n *such that* $n \cdot 3 = 0$, *the identity of* $\mathbb{Z}_7$.

$$
\begin{aligned}
&1 \cdot 3 = 3 \pmod 7 = 3 \\
&2 \cdot 3 = 3 \oplus_7 3 = (3+3) \pmod 7 = 6 \pmod 7 = 6 \\
&3 \cdot 3 = 3 \oplus_7 3 \oplus_7 3 = (3+3+3) \pmod 7 = 9 \pmod 7 = 2 \\
&4 \cdot 3 = 3 \oplus_7 3 \oplus_7 3 \oplus_7 3 = (3+3+3+3) \pmod 7 = 12 \pmod 7 = 5 \\
&5 \cdot 3 = 3 \oplus_7 3 \oplus_7 3 \oplus_7 3 \oplus_7 3 = (3+3+3+3+3) \pmod 7 = 15 \pmod 7 \\
&\qquad = 1 \\
&6 \cdot 3 = 3 \oplus_7 3 \oplus_7 3 \oplus_7 3 \oplus_7 3 \oplus_7 3 = (3+3+3+3+3+3) \pmod 7 = 18 \pmod 7 \\
&\qquad = 4 \\
&7 \cdot 3 = 3 \oplus_7 3 \oplus_7 3 \oplus_7 3 \oplus_7 3 \oplus_7 3 \oplus_7 3 = (3+3+3+3+3+3+3) \pmod 7 \\
&\qquad = 21 \pmod 7 = 0
\end{aligned}
$$

In determining the order of an element from a group, all that is needed is knowing how to combine elements from that group. The group operation could be presented by a Cayley table. Table 5.9 is the Cayley table for the group $H = \{\alpha, \beta, \gamma, \delta\}$ with operation $*$.

TABLE 5.9
Cayley Table for H

$*$	α	β	γ	δ
α	α	β	γ	δ
β	β	γ	δ	α
γ	γ	δ	α	β
δ	δ	α	β	γ

For H, $\mid \alpha \mid = 1$, as $\alpha^1 = \alpha = e$, the identity element for this group. The following calculations show that $\mid \beta \mid = 4$:

- $\beta^2 = \beta * \beta = \gamma \neq e$.
- $\beta^3 = \beta * \beta * \beta = (\beta * \beta) * \beta = \gamma * \beta = \delta \neq e$.
- $\beta^4 = \beta * \beta * \beta * \beta = (\beta * \beta) * (\beta * \beta) = \gamma * \gamma = \alpha = e$.

Exercises 5.3

1. Given the following permutations from S_{10}, perform the computations indicated.

   (a) For $\alpha = \begin{pmatrix} 0 & 1 & 2 & 3 & 4 & 5 & 6 & 7 & 8 & 9 \\ 3 & 2 & 1 & 6 & 5 & 7 & 9 & 8 & 4 & 0 \end{pmatrix}$, compute α^3.

   (b) For $\tau = \begin{pmatrix} 0 & 1 & 2 & 3 & 4 & 5 & 6 & 7 & 8 & 9 \\ 1 & 2 & 0 & 4 & 5 & 3 & 7 & 8 & 6 & 9 \end{pmatrix}$, compute τ^6.

   (c) For $\delta = (0,9)(1,2,3,8,7)(4,5)(6)$, compute δ^2.

   (d) For $\sigma = (0)(1,4)(2,3)(5,6,7,8,9)$, compute σ^1, σ^2, σ^3, σ^4, σ^5, σ^6, σ^7, σ^8, σ^9, and σ^{10}.

2. Consider the group $\mathbb{Z}_9$ with the operation $\oplus_9$. Perform the indicated computations. The group element from $\mathbb{Z}_9$ is underlined.

   (a) $1 \cdot \underline{3}$ (b) $4 \cdot \underline{5}$ (c) $7 \cdot \underline{2}$ (d) $9 \cdot \underline{3}$

3. Consider this Cayley table for the group $G = \{\bigstar, \sharp, \P, \S, \dagger, \exists\}$ with the operation $*$.

$*$	$\bigstar$	$\sharp$	$\P$	$\S$	$\dagger$	$\exists$
$\bigstar$	$\bigstar$	$\sharp$	$\P$	$\S$	$\dagger$	$\exists$
$\sharp$	$\sharp$	$\P$	$\bigstar$	$\dagger$	$\exists$	$\S$
$\P$	$\P$	$\bigstar$	$\sharp$	$\exists$	$\S$	$\dagger$
$\S$	$\S$	$\exists$	$\dagger$	$\bigstar$	$\P$	$\sharp$
$\dagger$	$\dagger$	$\S$	$\exists$	$\sharp$	$\bigstar$	$\P$
$\exists$	$\exists$	$\dagger$	$\S$	$\P$	$\sharp$	$\bigstar$

(a) Perform these calculations.

$$\bigstar^3 \qquad \sharp^2 \qquad \S^5 \qquad \dagger^9$$

(b) Find the order of each element from this group.

4. Find the order of each of the following elements from S_6.

(a) $f = \begin{pmatrix} 0 & 1 & 2 & 3 & 4 & 5 \\ 5 & 4 & 3 & 2 & 1 & 0 \end{pmatrix}$

(b) $g = \begin{pmatrix} 0 & 1 & 2 & 3 & 4 & 5 \\ 1 & 0 & 3 & 2 & 5 & 4 \end{pmatrix}$

(c) $h = \begin{pmatrix} 0 & 1 & 2 & 3 & 4 & 5 \\ 3 & 1 & 5 & 4 & 0 & 2 \end{pmatrix}$

5. Find the order of each of the elements from $\mathbb{Z}_{12}$.

6. In Exercise 5.3.3, you found the order of each element of $G = \{\bigstar, \sharp, \P, \S, \dagger, \exists\}$, and in Exercise 5.3.5, you found the order of each element in $\mathbb{Z}_{12}$. Relate the order of each group element to the order of the group it comes from. Do you notice any relationship? Make a conjecture about the order of any group element g from an arbitrary finite group G.

7. Let a be an element in a group G. If $| a |= 5$ (the order of a is 5), find a^5, a^6, a^7, a^8, a^9, a^{10}, a^{11}, a^{24}, and a^{100} in terms of e, a, a^2, a^3, and a^4. Recall that $| a |= 5$ means that $a^5 = e$.

Paper Assignment 5.3

1. **Summarizing.** Write a short summary of Section 5.3. Among other terms, your summary should include the following, but not necessarily in this order: *exponent, group, group element, power, operation*, and *order*. The explanation of these concepts should not appear as a list, but should be connected, such that your summary takes the form of an essay. Where notation is used, follow the format from the text.

Group Activities 5.3

1. In Exercise 5.3.3, you found the order of each element of $G = \{\bigstar, \sharp, \P, \S, \dagger, \exists\}$, and in Exercise 5.3.5, you found the order of each element in $\mathbb{Z}_{12}$.

   (a) List each of the elements from $G = \{\bigstar, \sharp, \P, \S, \dagger, \exists\}$ and the order of each of these elements. Divide the order obtained for each group element into the order of the group G ($| G |= 6$).

   (b) List each of the elements from $\mathbb{Z}_{12}$ and the order of each of these elements. Divide the order obtained for each group element into the order of $\mathbb{Z}_{12}$ ($| \mathbb{Z}_{12} |= 12$).

(c) Given the results of the work done in parts (a) and (b), make a conjecture about the order of each element in a particular group. In other words, if G is an arbitrary group with order n ($|G| = n$) and g is an element in G with order k (where $g^k = e$, and e is the group identity), what is the relation between k and n?

(d) Can there be an element of order 7 in a group of order 20, of order 25, or of order 11? Using the answer to this question and the result from part (c), what are all the possible group element orders in a group with order 20? A group of order 25? A group of order 11?

(e) Using the results from parts (c) and (d), what are the possible group element orders for a group of prime order, that is, a group G where the order of G is p, a prime number?

2. (a) Consider the group $\mathbb{Z}_{10} = \{0, 1, 2, 3, 4, 5, 6, 7, 8, 9\}$ and the group element 2 in $\mathbb{Z}_{10}$. This group element 2 will be underlined for the rest of this exercise. Compute $1 \cdot \underline{2}$, $2 \cdot \underline{2}$, $3 \cdot \underline{2}$, $4 \cdot \underline{2}$, and $5 \cdot \underline{2}$. Since $5 \cdot \underline{2} = 0$, the order of $\underline{2}$ is 5.

   (b) Now compute $6 \cdot \underline{2}$, $7 \cdot \underline{2}$, $8 \cdot \underline{2}$, $9 \cdot \underline{2}$, and $10 \cdot \underline{2}$ ($10 \cdot \underline{2}$ should equal 0). Is there pattern between these calculations and the ones performed in part (a)? Describe the pattern, if any.

   (c) Given the results from part (b), quickly find $11 \cdot \underline{2}$, $12 \cdot \underline{2}$, $13 \cdot \underline{2}$, $14 \cdot \underline{2}$, and $15 \cdot \underline{2}$. What do 1, 6, and 11 have in common? (HINT: Think about division by 5.) Repeat this process for 2, 7, and 12; for 3, 8, and 13; for 4, 9, and 14; and for 5, 10, and 15.

   (d) Based on your observations from part (c), quickly find $30 \cdot \underline{2}$, $101 \cdot \underline{2}$ and $214 \cdot \underline{2}$.

   (e) Now let a be an arbitrary element in a group G, where $|a| = 7$ (the order of a is 7). Raising a to powers results in the following list: $a^1 = a, a^2, a^3, a^4, a^5, a^6$; and $a^7 = e$.

   (f) Generalize the work you did in parts (a)-(d) and find a^8, a^9, a^{10}, a^{11}, a^{12}, a^{13}, and a^{14}, writing each in terms of the elements listed in part (e). Is there a pattern between these calculations and the ones performed in part (e)? Describe the pattern, if any.

   (g) Given the results from part (f), compute a^{15}, a^{16}, a^{17}, a^{18}, a^{19}, a^{20}, and a^{21}. What do 1, 8, and 15 have in common? (HINT: Think about division by 7.) Repeat this process for 2, 9, and 16; for 3, 10, and 17; for 4, 11, and 18; for 5, 12, and 19; for 6, 13, and 20; and for 7, 14, and 21.

   (h) Based on your observations from part (g), quickly find a^{24}, a^{28}, a^{71}, and a^{100}.

Further Reading

Hannabuss, K., Sound and Symmetry, *Mathematical Intelligencer*, 19(4), 1997, 16–21.

Humphreys, J. F., *A Course in Group Theory,* Oxford University Press, New York City, 1997.

White, A. T., Fabian Stedman: The First Group Theorist?, *American Mathematical Monthly*, 103(9), 1996, 771–778.

5.4 The Verhoeff Check Digit Scheme

Preliminary Activity. Most of the check digit schemes presented so far do not catch all transposition-of-adjacent-digits errors. One main reason is the use of addition, which is commutative, in these schemes. For example, recall the scheme presented in Exercise 2 of Chapter 1:

> Suppose you are an archivist at the Library of Congress and you have developed an identification number system to identify all of the different documents in the archives. Each document has a nine-digit $a_1a_2a_3a_4a_5a_6a_7a_8a_9$ identification number associated with it. The first seven digits $a_1a_2a_3a_4a_5a_6a_7$ identify the specific document, and the last two digits a_8 and a_9 are the check digits. For each document, the check digits a_8 and a_9 are determined by the sum of the digits in the document number. First calculate the sum $a_1+a_2+a_3+a_4+a_5+a_6+a_7$, and then assign a_8 to be the first (tens) digit and a_9 to be the second (ones) digit of this sum. In other words, the formula is $a_8a_9 = a_1+a_2+a_3+a_4+a_5+a_6+a_7$.
>
> For example, if the *Bill of Rights* has a document identification number of 2980162, then $a_8 = 2$ and $a_9 = 8$, since $2+9+8+0+1+6+2 = 28$. Consequently, the identification number associated with the *Bill of Rights* would be 298016228.
>
> If the sum of the first seven digits $(a_1+a_2+a_3+a_4+a_5+a_6+a_7)$ is a single digit a, then $a_8 = 0$ and $a_9 = a$. For example, consider the document number 1003201. Since the sum of the digits $1+0+0+3+2+0+1 = 7$, $a_8 = 0$, $a_9 = 7$, and the identification number would be 100320107.
>
> Consider the identification number 927361542. This would be an invalid number, as the sum of the first seven digits is $9+2+7+3+6+1+5 = 33$, which is not equal to the last two digits of 42. On the other hand, the identification number 927361533 would be a valid number.

This scheme will not catch transposition errors that do not involve the check digits. For example, if the number 701345222 was transmitted as $70\underline{31}45222$, the error would not be caught. The number 701345222 is valid since

$$7+0+1+3+4+5+2 = 22.$$

The error in the invalid number $70\underline{31}45222$ is not caught, since

$$7+0+3+1+4+5+2 = 22.$$

This happens because $1+3 = 3+1$.

With this problem in mind, develop a new way to combine the digits 0, 1, 2, 3, 4, 5, 6, 7, 8, and 9 that does not involve addition or any of the other operations from arithmetic (subtraction, multiplication, and division). Furthermore, make sure that this new method, denoted by $*$, is not commutative. In other words, $a * b$ should not equal $b * a$ for all digits a and b.

To this point, a variety of check digit schemes have been examined. As a general rule, a check digit scheme's goal is to catch as many as possible of the errors listed in Table 1.2. At the very least, the goal of a scheme should be to catch all single-digit and transposition-of-adjacent-digits errors, since they account for about 90% of all errors. Many schemes, like the US postal money order and the airline ticket schemes, did not even catch all single-digit errors. The UPC and IBM schemes caught all single-digit errors, but not all transposition-of-adjacent-digits errors.

One scheme did catch all single-digit and transposition-of-adjacent-digits errors. It was the ISBN scheme. However, it had two drawbacks. To work, the scheme introduced a new character, an X, into the identification number as a check digit. In other words, the check digit could be a $0, 1, 2, 3, 4, 5, 6, 7, 8, 9$, or X. The other drawback was that the ISBN scheme works only for identification numbers that are of length 10 (have ten digits in them). At that point, the new goal was stated: to develop a check digit scheme that at the very least catches all single-digit and transposition-of-adjacent-digits errors, uses only the digits 0 through 9 as check digits, and can be used with identification numbers of any length.

In 1969, J. Verhoeff [26] developed a check digit scheme that meets this goal and more. It not only catches all single-digit and transposition-of-adjacent-digits errors, but catches all of the error types listed in Table 1.2. It is also flexible, as it can be used with an identification number of any length. Information on the development and use of this scheme can be found in [3], [6], [7], [10], [11], and [29].

The Verhoeff scheme is based on two fundamental ideas.

1. Recall that in all the previous schemes studied, regular addition and multiplication were used in all the calculations for finding the check digit. Then modulo arithmetic was used to assign the check digit. The Verhoeff scheme uses a different method for combining digits.

   In Exercise 5 from Section 5.2, a Cayley table was created for the group D_{10} (the collection of symmetries of a regular 5-gon). First each of the ten elements was denoted by a digit from 0 to 9:

$$0 = \begin{pmatrix} a & b & c & d & e \\ a & b & c & d & e \end{pmatrix}, \quad 1 = \begin{pmatrix} a & b & c & d & e \\ b & c & d & e & a \end{pmatrix}, \quad 2 = \begin{pmatrix} a & b & c & d & e \\ c & d & e & a & b \end{pmatrix},$$

$$3 = \begin{pmatrix} a & b & c & d & e \\ d & e & a & b & c \end{pmatrix}, \quad 4 = \begin{pmatrix} a & b & c & d & e \\ e & a & b & c & d \end{pmatrix}, \quad 5 = \begin{pmatrix} a & b & c & d & e \\ a & e & d & c & b \end{pmatrix},$$

$$6 = \begin{pmatrix} a & b & c & d & e \\ e & d & c & b & a \end{pmatrix}, \quad 7 = \begin{pmatrix} a & b & c & d & e \\ d & c & b & a & e \end{pmatrix}, \quad 8 = \begin{pmatrix} a & b & c & d & e \\ c & b & a & e & d \end{pmatrix},$$

$$\text{and} \qquad 9 = \begin{pmatrix} a & b & c & d & e \\ b & a & e & d & c \end{pmatrix}.$$

TABLE 5.10
Cayley Table for D_{10}

*	0	1	2	3	4	5	6	7	8	9
0	0	1	2	3	4	5	6	7	8	9
1	1	2	3	4	0	6	7	8	9	5
2	2	3	4	0	1	7	8	9	5	6
3	3	4	0	1	2	8	9	5	6	7
4	4	0	1	2	3	9	5	6	7	8
5	5	9	8	7	6	0	4	3	2	1
6	6	5	9	8	7	1	0	4	3	2
7	7	6	5	9	8	2	1	0	4	3
8	8	7	6	5	9	3	2	1	0	4
9	9	8	7	6	5	4	3	2	1	0

The resulting Cayley table for D_{10}, presented in Table 5.10 with operation $*$, provides a new way to combine the digits 0 through 9 that is *not commutative*.

In check digit calculations for the other schemes studied, digits were multiplied or added together. For example, $3{\cdot}4 = 12$ or $3 + 4 = 7$ would be calculated. In the Verhoeff scheme, digits are combined using $*$, the results of which are presented in the Cayley table (Table 5.10) for D_{10}. For example $3 * 4 = 2$ and $8 * 3 = 5$.

2. In certain schemes, like the ISBN scheme, each digit from the identification number was multiplied by another number. For the ISBN scheme, the first digit was multiplied by 10, the second by 9, and so on. With the Verhoeff scheme, each digit a from the identification number is permuted by a permutation σ from S_{10} or by some power t of the permutation σ.

   In other words, instead of multiplying a by a number, we compute $\sigma^t(a)$, where t is a positive integer. Here, the permutation

$$\sigma = (0)(1,4)(2,3)(5,6,7,8,9),$$

   developed by S. Winters [29], is used.

Definition 5.4.1. The Verhoeff Check Digit Scheme. *Let $a_1a_2\ldots a_{n-1}a_n$ be an identification number with check digit a_n. The check digit a_n is appended to the number $a_1a_2\ldots a_{n-1}$ such that the following equation is satisfied:*

$$\sigma^{n-1}(a_1) * \sigma^{n-2}(a_2) * \sigma^{n-3}(a_3) * \cdots * \sigma(a_{n-1}) * a_n = 0$$

where $\sigma = (0)(1,4)(2,3)(5,6,7,8,9)$ and $*$ is the group operation from D_{10} as presented in Table 5.10.

Example 5.4.2. *The identification number 3170092 is a valid number using the Verhoeff scheme.*

- The number of digits in 3170092 is $n = 7$, where $a_1 = 3$, $a_2 = 1$, $a_3 = 7$, $a_4 = 0$, $a_5 = 0$, $a_6 = 9$, and the check digit $a_7 = 2$.
- For this number to be valid, the following calculation must be true.

$$
\begin{aligned}
\sigma^{n-1}(a_1) * \sigma^{n-2}(a_2) * \sigma^{n-3}(a_3) * \cdots * \sigma(a_{n-1}) * a_n &= 0 \\
\sigma^{7-1}(a_1) * \sigma^{7-2}(a_2) * \sigma^{7-3}(a_3) * \sigma^{7-4}(a_4) * \sigma^{7-5}(a_5) * \sigma^{7-6}(a_6) * a_7 &= 0 \\
\sigma^{6}(a_1) * \sigma^{5}(a_2) * \sigma^{4}(a_3) * \sigma^{3}(a_4) * \sigma^{2}(a_5) * \sigma^{1}(a_6) * a_7 &= 0 \\
\sigma^{6}(3) * \sigma^{5}(1) * \sigma^{4}(7) * \sigma^{3}(0) * \sigma^{2}(0) * \sigma(9) * 2 &= 0 \\
3 * 4 * 6 * 0 * 0 * 5 * 2 &= 0 \\
(3 * 4) * (6 * 0) * (0 * 5) * 2 &= 0 \\
2 * 6 * 5 * 2 &= 0 \\
(2 * 6) * (5 * 2) &= 0 \\
8 * 8 &= 0 \\
0 &= 0
\end{aligned}
$$

- Since $0 = 0$ is a true statement, 3170092 is valid.

Example 5.4.3. *The identification number* 57014 *is an invalid number using the Verhoeff scheme.*

- The number of digits in 57014 is $n = 5$, where $a_1 = 5$, $a_2 = 7$, $a_3 = 0$, $a_4 = 1$, and the check digit $a_5 = 4$.
- For this number to be invalid, the following calculation must be false.

$$
\begin{aligned}
\sigma^{n-1}(a_1) * \sigma^{n-2}(a_2) * \sigma^{n-3}(a_3) * \cdots * \sigma(a_{n-1}) * a_n &= 0 \\
\sigma^{5-1}(a_1) * \sigma^{5-2}(a_2) * \sigma^{5-3}(a_3) * \sigma^{5-4}(a_4) * a_5 &= 0 \\
\sigma^{4}(a_1) * \sigma^{3}(a_2) * \sigma^{2}(a_3) * \sigma^{1}(a_4) * a_5 &= 0 \\
\sigma^{4}(5) * \sigma^{3}(7) * \sigma^{2}(0) * \sigma(1) * 4 &= 0 \\
9 * 5 * 0 * 4 * 4 &= 0 \\
(9 * 5) * (0 * 4) * 4 &= 0 \\
4 * 4 * 4 &= 0 \\
(4 * 4) * 4 &= 0 \\
3 * 4 &= 0 \\
2 &= 0
\end{aligned}
$$

- Since $2 = 0$ is a false statement, 57014 is invalid.

Example 5.4.4. *Consider the five-digit number* 27163 *that has been generated to identify a certain item. The Verhoeff scheme will be used to append a check digit K to the number* 27163 *to create the identification number* $27163K$.

- This will be a six-digit number ($n = 6$) with $a_1 = 2$, $a_2 = 7$, $a_3 = 1$, $a_4 = 6$, $a_5 = 3$, and the check digit $a_6 = K$.

- The check digit K must satisfy the equation

$$\sigma^{n-1}(a_1) * \sigma^{n-2}(a_2) * \sigma^{n-3}(a_3) * \cdots * \sigma(a_{n-1}) * a_n = 0.$$

That is,

$$\sigma^{6-1}(a_1) * \sigma^{6-2}(a_2) * \sigma^{6-3}(a_3) * \sigma^{6-4}(a_4) * \sigma^{6-5}(a_5) * a_6 = 0,$$

or

$$\sigma^5(a_1) * \sigma^4(a_2) * \sigma^3(a_3) * \sigma^2(a_4) * \sigma^1(a_5) * a_6 = 0.$$

- In this case,

$$\sigma^5(2) * \sigma^4(7) * \sigma^3(1) * \sigma^2(6) * \sigma(3) * K = 0.$$

- To find the check digit K, solve this equation.

$$\begin{aligned}
\sigma^5(2) * \sigma^4(7) * \sigma^3(1) * \sigma^2(6) * \sigma(3) * K &= 0 \\
3 * 6 * 4 * 8 * 2 * K &= 0 \\
(3 * 6) * (4 * 8) * 2 * K &= 0 \\
9 * 7 * 2 * K &= 0 \\
(9 * 7) * 2 * K &= 0 \\
2 * 2 * K &= 0 \\
(2 * 2) * K &= 0 \\
4 * K &= 0
\end{aligned}$$

- Now use the Cayley table for D_{10} to find the digit K such that $4 * K = 0$. The only possible solution is $K = 1$.
- Thus the identification number is 271631.

As was mentioned earlier, the Verhoeff scheme can be used with identification numbers of any length.

Example 5.4.5. *Suppose a company has a* 12*-digit identification number system (an* 11*-digit product number plus a check digit). The company generates the* 11*-digit product number* 38601842927 *and wants to assign, using the Verhoeff scheme, a check digit* K *to create the identification number* $38601842927K$.

- The number of digits n in the identification number is 12 ($n = 12$), where $a_1 = 3$, $a_2 = 8$, $a_3 = 6$, $a_4 = 0$, $a_5 = 1$, $a_6 = 8$, $a_7 = 4$, $a_8 = 2$, $a_9 = 9$, $a_{10} = 2$, $a_{11} = 7$, and the check digit $a_{12} = K$.
- The check digit K must satisfy the equation

$$\sigma^{11}(3) * \sigma^{10}(8) * \sigma^9(6) * \sigma^8(0) * \sigma^7(1) * \sigma^6(8) * \sigma^5(4) * \sigma^4(2) * \sigma^3(9) \\ * \sigma^2(2) * \sigma(7) * K = 0$$

- To find the check digit K, solve this equation.

$$\begin{aligned}\sigma^{11}(3) * \sigma^{10}(8) * \sigma^{9}(6) * \sigma^{8}(0) * \sigma^{7}(1) * \sigma^{6}(8) * \sigma^{5}(4) * \sigma^{4}(2)&\\ *\sigma^{3}(9) * \sigma^{2}(2) * \sigma(7) * K &= 0\\ 2 * 8 * 5 * 0 * 4 * 9 * 1 * 2 * 7 * 2 * 8 * K &= 0\\ (2 * 8) * (5 * 0) * (4 * 9) * (1 * 2) * (7 * 2) * 8 * K &= 0\\ 5 * 5 * 8 * 3 * 5 * 8 * K &= 0\\ (5 * 5) * (8 * 3) * (5 * 8) * K &= 0\\ 0 * 5 * 2 * K &= 0\\ (0 * 5) * 2 * K &= 0\\ 5 * 2 * K &= 0\\ (5 * 2) * K &= 0\\ 8 * K &= 0\end{aligned}$$

- Now use the Cayley table for D_{10} to find the digit K such that $8 * K = 0$. The only possible solution is $K = 8$.
- Thus the identification number is 386018429278.

As mentioned above, the Verhoeff scheme catches all single-digit and transposition-of-adjacent-digits errors.

Example 5.4.6. *If the correct number* 38601<u>8</u>429278 *incurred a single-digit error and was transmitted as the number* 38601<u>5</u>429278*, it would be caught by the Verhoeff scheme.*

- Compute the calculation

$$\begin{aligned}\sigma^{11}(3) * \sigma^{10}(8) * \sigma^{9}(6) * \sigma^{8}(0) * \sigma^{7}(1) * \sigma^{6}(5) * \sigma^{5}(4) * \sigma^{4}(2) * \sigma^{3}(9)&\\ * \sigma^{2}(2) * \sigma(7) * 8 = 0\ .&\end{aligned}$$

- This leads to a false statement:

$$\begin{aligned}\sigma^{11}(3) * \sigma^{10}(8) * \sigma^{9}(6) * \sigma^{8}(0) * \sigma^{7}(1) * \sigma^{6}(5) * \sigma^{5}(4) * \sigma^{4}(2)&\\ *\sigma^{3}(9) * \sigma^{2}(2) * \sigma(7) * 8 &= 0\\ 2 * 8 * 5 * 0 * 4 * 6 * 1 * 2 * 7 * 2 * 8 * 8 &= 0\\ (2 * 8) * (5 * 0) * (4 * 6) * (1 * 2) * (7 * 2) * (8 * 8) &= 0\\ 5 * 5 * 5 * 3 * 5 * 0 &= 0\\ (5 * 5) * (5 * 3) * (5 * 0) &= 0\\ 0 * 7 * 5 &= 0\\ (0 * 7) * 5 &= 0\\ 7 * 5 &= 0\\ 2 &= 0.\end{aligned}$$

- Since $2 = 0$ is a false statement, this single-digit error would be caught.

It is very easy to show why the Verhoeff scheme catches all single-digit errors. Let $a_1 \dots a_i \dots a_n$ be an identification number generated using the Verhoeff check digit scheme. When a single-digit error occurs, $a_1 \dots a_i \dots a_n$ is transmitted as $a_1 \dots b_i \dots a_n$ with $a_i \neq b_i$ (this is a single-digit error where a_i is replaced by b_i).

Suppose this error is not caught. Then applying the Verhoeff check digit scheme calculation to both the correct and the incorrect numbers will result in a true statement in each case. This results in

$$\sigma^{n-1}(a_1) * \cdots * \sigma^{n-i}(a_i) * \cdots * a_n = 0$$

and

$$\sigma^{n-1}(a_1) * \cdots * \sigma^{n-i}(b_i) * \cdots * a_n = 0.$$

Equivalently,

$$\sigma^{n-1}(a_1) * \cdots * \sigma^{n-i}(a_i) * \cdots * a_n = \sigma^{n-1}(a_1) * \cdots * \sigma^{n-i}(b_i) * \cdots * a_n.$$

Since the expression

$$\sigma^{n-1}(a_1) * \cdots * \sigma^{n-i}(a_i) * \cdots * a_n$$

is identical to the expression

$$\sigma^{n-1}(a_1) * \cdots * \sigma^{n-i}(b_i) * \cdots * a_n,$$

except for the terms $\sigma^{n-i}(a_i)$ and $\sigma^{n-i}(b_i)$, the only way for these two entire expressions to be equal is when $\sigma^{n-i}(a_i) = \sigma^{n-i}(b_i)$. Thus

$$\sigma^{n-i}(a_i) = \sigma^{n-i}(b_i).$$

However, since σ^{n-i} is a permutation (one-to-one and onto) in S_{10}, the only way for $\sigma^{n-i}(a_i)$ to equal $\sigma^{n-i}(b_i)$ is if $a_i = b_i$. This is a contradiction. Thus the assumption that error is not caught is false, and the single-digit error is caught.

In addition, the Verhoeff scheme will catch all transposition-of-adjacent-digits errors.

Example 5.4.7. *Suppose that the correct number* $3860184\underline{29}278$ *incurred a transposition-of-adjacent-digits error and was transmitted as the number* $3860184\underline{92}278$. *This would be caught by the Verhoeff scheme.*

- Compute the calculation

$$\sigma^{11}(3) * \sigma^{10}(8) * \sigma^{9}(6) * \sigma^{8}(0) * \sigma^{7}(1) * \sigma^{6}(8) * \sigma^{5}(4) * \sigma^{4}(9) * \sigma^{3}(2) \\ * \sigma^{2}(2) * \sigma(7) * 8 = 0.$$

- This leads to a false statement:

$$\begin{aligned} \sigma^{11}(3) * \sigma^{10}(8) * \sigma^{9}(6) * \sigma^{8}(0) * \sigma^{7}(1) * \sigma^{6}(8) * \sigma^{5}(4) * \sigma^{4}(9) \\ * \sigma^{3}(2) * \sigma^{2}(2) * \sigma(7) * 8 &= 0 \\ 2 * 8 * 5 * 0 * 4 * 9 * 1 * 8 * 3 * 2 * 8 * 8 &= 0 \\ (2 * 8) * (5 * 0) * (4 * 9) * (1 * 8) * (3 * 2) * (8 * 8) &= 0 \\ 5 * 5 * 8 * 9 * 0 * 0 &= 0 \\ (5 * 5) * (8 * 9) * (0 * 0) &= 0 \\ 0 * 4 * 0 &= 0 \\ (0 * 4) * 0 &= 0 \\ 4 * 0 &= 0 \\ 4 &= 0. \end{aligned}$$

- Since $4 = 0$ is a false statement, this transposition-of-adjacent-digits error would be caught.

German Bundesbank Check Digit Scheme In 1990, the German Bundesbank (Federal Bank) began using a scheme based on the Verhoeff check digit scheme [6]. While it does use the familiar Cayley table for D_{10} (Table 5.11), the Bundesbank's variation has two differences:

1. It uses the permutation $\sigma = (0, 1, 5, 8, 9, 4, 2, 7)(3, 6)$ from S_{10} instead of the permutation $\sigma = (0)(1, 4)(2, 3)(5, 6, 7, 8, 9)$.
2. It uses ascending powers of σ instead of descending powers as the Verhoeff scheme does.

TABLE 5.11
Cayley Table for D_{10}

*	0	1	2	3	4	5	6	7	8	9
0	0	1	2	3	4	5	6	7	8	9
1	1	2	3	4	0	6	7	8	9	5
2	2	3	4	0	1	7	8	9	5	6
3	3	4	0	1	2	8	9	5	6	7
4	4	0	1	2	3	9	5	6	7	8
5	5	9	8	7	6	0	4	3	2	1
6	6	5	9	8	7	1	0	4	3	2
7	7	6	5	9	8	2	1	0	4	3
8	8	7	6	5	9	3	2	1	0	4
9	9	8	7	6	5	4	3	2	1	0

FIGURE 5.2
German Bank Note with Serial Number GK7042314S and Check Digit 5

The identification numbers on German bank notes are an 11-character alphanumeric system. An example is given in Figure 5.2.

To use the Verhoeff scheme with the permutation σ mentioned above, the letters that appear in the identification number need to be assigned numerical values. The method used is presented in Table 5.12.

TABLE 5.12
Numerical Values Used by the Bundesbank

A	D	G	K	L	N	S	U	Y	Z
0	1	2	3	4	5	6	7	8	9

Definition 5.4.8. The German Bundesbank Check Digit Scheme. *Consider the identification number* $a_1a_2a_3a_4a_5a_6a_7a_8a_9a_{10}a_{11}$ *with check digit* a_{11}. *The check digit* a_{11} *is appended to the number* $a_1a_2a_3a_4a_5a_6a_7a_8a_9a_{10}$ *such that the following equation is satisfied:*

$$\sigma(a_1) * \sigma^2(a_2) * \sigma^3(a_3) * \sigma^4(a_4) * \sigma^5(a_5) * \sigma^6(a_6) * \sigma^7(a_7) * \sigma^8(a_8)$$
$$* \sigma^9(a_9) * \sigma^{10}(a_{10}) * a_{11} = 0$$

where $\sigma = (0, 1, 5, 8, 9, 4, 2, 7)(3, 6)$ *and* $*$ *is the group operation from* D_{10} *as presented in Table 5.11.*

The identification number GK7042314S5 is a valid number. To show this, the number must first be rewritten using the numerical values presented in Table 5.12. Since G $\to$ 2,

K → 3, and S → 6, the numerical equivalent of GK7042314S5 is 23704231465. The following calculation results in a true statement, so the number is valid.

$$\begin{aligned}\sigma(a_1) * \sigma^2(a_2) * \sigma^3(a_3) * \sigma^4(a_4) * \sigma^5(a_5) * \sigma^6(a_6) * \sigma^7(a_7)&\\ *\sigma^8(a_8) * \sigma^9(a_9) * \sigma^{10}(a_{10}) * a_{11} &= 0\\ \sigma(2) * \sigma^2(3) * \sigma^3(7) * \sigma^4(0) * \sigma^5(4) * \sigma^6(2) * \sigma^7(3)&\\ *\sigma^8(1) * \sigma^9(4) * \sigma^{10}(6) * 5 &= 0\\ 7 * 3 * 5 * 9 * 5 * 9 * 6 * 1 * 2 * 6 * 5 &= 0\\ (7 * 3) * (5 * 9) * (5 * 9) * (6 * 1) * (2 * 6) * 5 &= 0\\ 9 * 1 * 1 * 5 * 8 * 5 &= 0\\ (9 * 1) * (1 * 5) * (8 * 5) &= 0\\ 8 * 6 * 3 &= 0\\ (8 * 6) * 3 &= 0\\ 2 * 3 &= 0\\ 0 &= 0\end{aligned}$$

The German Bank note number AK7320671S3 would be invalid as the following calculation results in a false statement. (Note: The numerical equivalent of AK7320671S3 is 03732067163).

$$\begin{aligned}\sigma(a_1) * \sigma^2(a_2) * \sigma^3(a_3) * \sigma^4(a_4) * \sigma^5(a_5) * \sigma^6(a_6) * \sigma^7(a_7)&\\ *\sigma^8(a_8) * \sigma^9(a_9) * \sigma^{10}(a_{10}) * a_{11} &= 0\\ \sigma(0) * \sigma^2(3) * \sigma^3(7) * \sigma^4(3) * \sigma^5(2) * \sigma^6(0) * \sigma^7(6)&\\ *\sigma^8(7) * \sigma^9(1) * \sigma^{10}(6) * 3 &= 0\\ 1 * 3 * 5 * 3 * 8 * 2 * 3 * 7 * 5 * 6 * 3 &= 0\\ (1 * 3) * (5 * 3) * (8 * 2) * (3 * 7) * (5 * 6) * 3 &= 0\\ 4 * 7 * 6 * 5 * 4 * 3 &= 0\\ (4 * 7) * (6 * 5) * (4 * 3) &= 0\\ 6 * 1 * 2 &= 0\\ (6 * 1) * 2 &= 0\\ 5 * 2 &= 0\\ 8 &= 0\end{aligned}$$

Suppose the German Bundesbank has the serial number AN9941270L and wants to assign a check digit C to this number to create the identification number AN9941270LC. The same method is used again. First, using the numerical values, find the numerical equivalent of AN9941270LC. This is 0599412704C. Then C must be a digit that satisfies the equation

$$\sigma(a_1) * \sigma^2(a_2) * \sigma^3(a_3) * \sigma^4(a_4) * \sigma^5(a_5) * \sigma^6(a_6) * \sigma^7(a_7) * \sigma^8(a_8) * \sigma^9(a_9) * \sigma^{10}(a_{10}) * C = 0.$$

To find C, the following equation is solved:

$$\begin{aligned}
\sigma(0) * \sigma^2(5) * \sigma^3(9) * \sigma^4(9) * \sigma^5(4) * \sigma^6(1) * \sigma^7(2) * \sigma^8(7) \\
*\sigma^9(0) * \sigma^{10}(4) * C = 0 \\
1 * 9 * 7 * 0 * 5 * 7 * 4 * 7 * 1 * 7 * C = 0 \\
(1 * 9) * (7 * 0) * (5 * 7) * (4 * 7) * (1 * 7) * C = 0 \\
5 * 7 * 3 * 6 * 8 * C = 0 \\
(5 * 7) * (3 * 6) * 8 * C = 0 \\
3 * 9 * 8 * C = 0 \\
(3 * 9) * 8 * C = 0 \\
7 * 8 * C = 0 \\
(7 * 8) * C = 0 \\
4 * C = 0.
\end{aligned}$$

Looking at the table for D_{10} (Table 5.11), the solution to $4 * C = 0$ is $C = 1$. Thus the check digit is 1 and the identification number is AN9941270L1.

There is a slight problem with the German Bundesbank scheme. Although it is based on the Verhoeff scheme (which catches all the errors listed in Table 1.2), the Bundesbank scheme does not catch all single-digit errors. This problem occurs because of the numerical values assigned to the letters in the identification number. If a single-digit error changes a letter to its numerical value, that error will not be caught. For example, consider the valid number GK7042314S5. The numerical value of G is 2. If the single-digit error of recording $\underline{\text{G}}$K7042314S5 as $\underline{2}$K7042314S5 occurred, it would not be caught. The scheme also does not catch all transposition-of-adjacent-digits errors.

Final Comment. The advent of the information and technology age requires quick and reliable transmission of data and information. Methods such as check digit schemes ensure that this is done confidentially and without error. Today's computer-reliant society uses the material developed in this book on a daily basis in all walks of life. This is illustrated by the simple procedure of sending a credit card number over the Internet. Not only do credit card numbers use the IBM check digit scheme presented in Chapter 3, but the RSA public key cryptography system is used to send the number from the buyer, via the Internet, to the merchant.

One question still remains: Why do many identification number systems still use check digit schemes that do not catch all of the errors listed in Table 1.2? This is a good question for which I have no answer. However, it does motivate the need to adjust current schemes and to create new ones. All of the concepts introduced in this book can be applied to create sophisticated schemes that will catch most errors. This was demonstrated by the Verhoeff scheme. With the birth of new identification number systems and the existence of old and ineffective check digit schemes, innovative schemes must be developed to ensure that data and information continue to flow error-free.

Exercises 5.4

1. (a) High-Tech, a computer software company, has decided to use the check digit scheme designed by J. Verhoeff. Each identification number at High-Tech is a seven-digit number $a_1a_2a_3a_4a_5a_6a_7$, where $a_1a_2a_3a_4a_5a_6$ is the product number and a_7 is the check digit. Suppose that the following two numbers are High-Tech identification numbers. Using the Verhoeff check digit scheme, determine which is valid and which is invalid. Remember that the last digit is the check digit.

   3216903 9126774

   (b) The number 861492 will identify a High-Tech product. Use the Verhoeff check digit scheme to assign a check digit to this number. Be sure to identify the check digit and then to write out the entire identification number (including the check digit).

2. (a) Low-Tech, another computer software company, has also decided to use the check digit scheme designed by J. Verhoeff. Each identification number at Low-Tech is a ten-digit number $a_1a_2a_3a_4a_5a_6a_7a_8a_9a_{10}$, where $a_1a_2a_3a_4a_5a_6a_7a_8a_9$ is the product number and a_{10} is the check digit. Suppose that the following two numbers are Low-Tech identification numbers. Using the Verhoeff check digit scheme, determine which is valid and which is invalid. Remember that the last digit is the check digit.

   3216903549 9126774022

   (b) The number 861492553 will identify a Low-Tech product. Use the Verhoeff check digit scheme to assign a check digit to this number. Be sure to identify the check digit and then to write out the entire identification number (including the check digit).

3. (a) Consider the following two Bundesbank money identification numbers. Using the check digit scheme of the Bundesbank, determine which is valid and which is invalid.

   LS1503700Z3 DN3024798U4

   (b) The alphanumeric serial number AG8506827K will identify a bank note issued by the Bundesbank. Using the check digit scheme of the Bundesbank, assign a check digit to this serial number. Be sure to identify the check digit and then to write out the entire identification number (including the check digit).

4. What kind of transposition-of-adjacent-digits errors will elude the Bundesbank check digit scheme? Explain how you obtained your answer.

Paper Assignments 5.4

1. **Comparison.** You are employed by an organization that wants to use a check digit scheme. At this point, you are familiar with a variety of schemes. An initial survey

done by the company has reduced the choices to two schemes, the ISBN and Verhoeff schemes. Your colleague recommends the ISBN scheme, but you believe that the Verhoeff scheme is better for the organization. Write a report that not only recommends the Verhoeff scheme but also refutes your colleague's earlier recommendation. To do this, compare both schemes in a report called a feasibility study, paying attention to the following factors:

(a) ease of use (the length, the time involved with, and the complexity of the calculations);

(b) strength (in terms of the errors it can detect);

(c) flexibility (the ISBN scheme is designed for an identification number of a fixed length while the Verhoeff scheme can be adjusted to an identification number of any length).

At the end of this comparison, recommend the Verhoeff scheme and state your reasons for doing so.

2. **Comparison.** You are employed by an organization that wants to use a check digit scheme. At this point, you are familiar with a variety of schemes. An initial survey done by the company has reduced the choices to two schemes, the IBM and Verhoeff schemes. Your colleague recommends the IBM scheme, but you believe that the Verhoeff scheme is better for the organization. Write a report that not only recommends the Verhoeff scheme but also refutes your colleague's earlier recommendation. To do this, compare both schemes in a report called a feasibility study, paying attention to the following factors:

   (a) ease of use (the length, the time involved with, and the complexity of the calculations);

   (b) strength (in terms of the errors it can detect);

   (c) flexibility (both the IBM and the Verhoeff schemes are designed for an identification number of any length).

   At the end of this comparison, recommend the Verhoeff scheme and state your reasons for doing so.

Group Activities 5.4

1. As mentioned above, the German Bundesbank check digit scheme does not catch all transposition-of-adjacent-digits errors. If a and b are two adjacent digits in a German bank note number, the transposition error $\ldots ab \cdots \rightarrow \ldots ba \ldots$ is not always caught. The goal of this activity is to determine when a transposition-of-adjacent-digits error can be caught. For each of the following valid German bank note numbers, two transposition-of-adjacent-digits errors are shown. Each member of the group chooses one valid number to investigate.

Correct Number	Incorrect Number A	Incorrect Number B
LS6128132K6	LS6218132K6	L6S128132K6
KZ2136851D7	ZK2136851D7	KZ213685D17
DG2358115N0	DG2385115N0	D2G385115N0
DG2358115N0	DG2351815N0	DG235811N50

(a) Making sure to write out all the calculations, show that the Correct Number is actually a valid identification number.

(b) Making sure to write out all the calculations, show that the transposition error in Incorrect Number A will be caught, while the transposition error in Incorrect Number B will not be caught.

(c) In answering parts (a) and (b), the following was computed for each number where $a_1a_2a_3a_4a_5a_6a_7a_8a_9a_{10}a_{11}$ was the correct or incorrect number.

$$\sigma(a_1) * \sigma^2(a_2) * \sigma^3(a_3) * \sigma^4(a_4) * \sigma^5(a_5) * \sigma^6(a_6) * \sigma^7(a_7) * \sigma^8(a_8) \\ * \sigma^9(a_9) * \sigma^{10}(a_{10}) * a_{11} = 0.$$

Compare the calculation

$$\sigma(a_1) * \sigma^2(a_2) * \sigma^3(a_3) * \sigma^4(a_4) * \sigma^5(a_5) * \sigma^6(a_6) * \sigma^7(a_7) * \sigma^8(a_8) \\ * \sigma^9(a_9) * \sigma^{10}(a_{10}) * a_{11}$$

done for the Correct Number with Incorrect Numbers A and B. In both cases, there should be only a minor difference between the pair of calculations. What is it in each case? Might the permutation σ shed some light on your investigation? Explain.

(d) Given your work in the previous steps, conjecture as to when the transposition of adjacent digits a and b $(\ldots ab \ldots \to \ldots ba \ldots)$ is not caught.

Further Reading

Gallian, J., The Mathematics of Identification Numbers, *College Mathematics Journal*, 22(3), 1991, 194–202.

Gallian, J., Error Detection Methods, *ACM Computing Surveys*, 28(3), 1996, 504–517.

Winters, S. J., Error Detecting Schemes Using Dihedral Groups, *UMAP Journal*, 11(4), 1990, 299–308.

6

Bibliography

Armstrong, M. A. *Groups and Symmetry.* Springer-Verlag, New York City, 1988.

Beker, H., and Piper, F. *Cipher Systems: The Protection of Communications.* Wiley, New York City, 1982.

Bernard, K. J., and Wellenzohn, H. J. *Foundations of Mathematics.* H&H Publishing Company, Clearwater, FL, 1997.

Beutelspacher, A. *Cryptology.* The Mathematical Association of America, Washington, D.C., 1994.

Boneh, D., Twenty Years of Attacks on the RSA Cryptosystem. *Notices of the AMS,* 46(2), 1999, 203–213.

Briggs, J. *Fractals: The Patterns of Chaos.* Simon and Schuster, New York City, 1992.

Burton, D. M. *Elementary Number Theory.* McGraw Hill, New York City, 1998.

Burton, D. M. *The History of Mathematics.* WCB McGraw Hill, Boston, 1999.

D'Angelo, J. P., and West, D. W. *Mathematical Thinking: Problem Solving and Proofs.* Prentice Hall, Upper Saddle River, NJ, 1997.

Davis, D. *The Nature and Power of Mathematics.* Princeton University Press, Princeton, NJ, 1993.

Devlin, K., *Mathematics, The Science of Patterns: The Search for Order in Life, Mind, and the Universe.* W.H. Freeman, New York City, 1994.

Dubinsky, E., and Harvel, G., eds. *The Concept of Function: Aspects of Epistemology and Pedagogy.* MAA Notes, No. 25. The Mathematical Association of America, Washington, D.C., 1992.

Escher, M. C. *The Infinite World of M. C. Escher.* Abradale Press, New York City, 1984.

Escher, M. C. *Escher on Escher.* H.N. Abrams, New York City, 1989.

Field, M., and Golubitsky, M. *Symmetry in Chaos: A Search for Pattern in Mathematics, Art, and Nature.* Oxford University Press, Oxford, 1995.

Gallian, J. A. *Contemporary Abstract Algebra.* Houghton Mifflin, Boston, 1998.

Garliènski, J. *The Enigma War.* Scribner, New York City, 1979.

Gardner, M. *Time Travel and Other Mathematical Bewilderments.* W.H. Freeman, New York City, 1988.

Gardner, M. *The New Ambidestrous Universe: Symmetry and Asymmetry from Mirror Reflections to Superstrings.* W.H. Freeman, New York City, 1990.

Grünbaum, B., *Tilings and Patterns.* Freeman, New York City, 1987.

Halmos, P. R. *Naive Set Theory.* Springer-Verlag, New York City, 1974.

Hannabuss, K. Sound and Symmetry. *The Mathematical Intelligencer*, 19(4), 1997, 16–21.

Hargittai, I., ed. *Symmetry: Unifying Human Understanding.* Pergamon, New York City, 1986.

Harris, R. The Power of Information Graphics. *IIE Solutions*, 31, 1999, 26–27.

Herstein, I. N. *Abstract Algebra.* Prentice Hall, Upper Saddle River, NJ, 1990.

Humphreys, J. F. *Numbers, Groups, and Codes.* Cambridge University Press, Cambridge, 1989.

Humphreys, J. F. *A Course in Group Theory.* Oxford University Press, New York City, 1997.

Isihara, P., and Knapp, M. Basic Z_{12} Analysis of Musical Chords. *UMAP Journal*, 14(4), 1993, 319–348.

Johnston, B. L., and Richman, R. *Numbers and Symmetry: An Introduction to Algebra.* CRC Press, Boca Raton, 1997.

Jones, C. *Navajo Code Talkers: Native American Heroes.* Tudor Publishers, Greensboro, 1997.

Kawano, K., et al. *Warriors: Navajo Code Talkers.* Northland Publishing, Flagstaff, 1990.

Khan, D. *Seizing the Enigma: The Race to Break the German U-Boat Codes, 1939–1943.* Houghton Mifflin, Boston, 1991.

Kahn, D. *The Codebreakers: The Comprehensive History of Secret Communications from Ancient Times to the Internet.* Scribner, New York City, 1996.

Lee, B. *Marching Orders: The Untold Story of World War II.* Crown Publishers, New York City, 1995.

Lewin, R. *The American Magic: Codes, Ciphers, and the Defeat of Japan.* Farrar Straus Giroux, New York City, 1982.

Luciano, D., and Prichett, G. Cryptology: From Caesar Ciphers to Public-Key Cryptosystems. *College Mathematics Journal*, 18(1), 1987, 2–17.

Menezes, A. J., van Oorschot, P., and Vanstone, S. *Handbook of Applied Cryptography.* CRC Press, Boca Raton, 1996.

Mueller, A., et al. More Missouri Breaks. *UMAP Journal*, 13(4), 1992, 351–352.

Newton, D. E. *Encyclopedia of Cryptology.* ABC-CLIO, Santa Barbara, 1997.

Ogilvy C. S., and Anderson, J. T. *Excursions in Number Theory.* Dover, New York City, 1966.

Ore, O. *Invitation to Number Theory.* Random House, New York City, 1967.

Pedersen, F. D. *Modern Algebra: A Conceptual Approach.* W.C. Brown, Dubuque, Iowa, 1993.

Pinter, C. C. *A Book of Abstract Algebra.* McGraw Hill, New York City, 1990.

Pless, V., *Introduction to the Theory of Error-Correcting Codes.* Wiley, New York City, 1982.

Pomerance, C. The Search for Prime Numbers. *Scientific American*, 247, 1982, 122–130.

Prados, J. *Combined Fleet Decoded: The Secret History of American Intelligence and the Japanese Navy in World War II.* Random House, New York City, 1995.

Richardson, D., and St. John, P. Plotting to Succeed. *Geographical Magazine*, 64(12), 1992, 42–44.

Roberts, J. *Lure of the Integers.* The Mathematical Association of America, Washington, D.C., 1992.

Schattschneider, D. *Visions of Symmetry: Notebooks, Periodic Drawings, and Related Work of M. C. Escher.* W. H. Freeman, New York City, 1990.

Seberry, J., and Pieprzyk, J. *Cryptography: An Introduction to Computer Security.* Prentice Hall, Upper Saddle River, NJ, 1989.

Stewart, I. The Art of Elegant Tiling. *Scientific American*, 281(1), 1999, 96–98.

Stinson, D. R., *Cryptography: Theory and Practice.* CRC Press, Boca Raton, 1995.

UPC Symbol Specification Manual. Uniform Code Council, Dayton, Ohio, 1986.

Vinzant, C. What Hidden Meanings Are Embedded in Your Social Security Number? *Fortune*, 139, 1999, 32.

Welchman, G. *The Hut Six Story: Breaking the Enigma Codes.* McGraw Hill, New York City, 1982.

Wertenbaker, C. Nature's Patterns. *Parabola*, 24 1999, 5–12.

White, A. T. Fabian Stedman: The First Group Theorist? *American Mathematical Monthly*, 103(9), 1996, 771–778.

Wood, E. F. Self-Checking Codes: An Application of Modular Arithmetic. *Mathematics Teacher*, 80, 1987, 312–316.

Wrixon, F. B. *Codes and Ciphers.* Prentice Hall, Upper Saddle River, NJ, 1992.

Wussing, H. *The Genesis of the Abstract Group Concept.* MIT Press, Cambridge, 1984.

Zimmermann, R. R. Cryptography for the Internet. *Scientific American*, 279(4), 1998, 110–115.

References Cited

[1] Brown, D. A. H., Construction of Error Detection and Error Correction Codes to Any Base. *Electronic Letters*, 9, 1973, 290.

[2] Durbin, J .R., *Modern Algebra: An Introduction.* Wiley, New York City, 1992.

[3] Gallian, J. A., Check Digit Methods. *International Journal of Applied Engineering Education*, 5(4), 1989, 503–505.

[4] Gallian, J. A., The Mathematics of Identification Numbers. *College Mathematics Journal*, 22(3), 1991, 194–202.

[5] Gallian, J. A., Assigning Driver's License Numbers. *Mathematics Magazine*, 64(1), 1991, 13–22.

[6] Gallian, J. A., Error Detection Methods. *ACM Computing Surveys*, 28(3), 1996, 504–517.

[7] Gallian, J. A., and Winters, S., Modular Arithmetic in the Marketplace. *American Mathematical Monthly*, 95, 1988, 548–551.

[8] Garliènski, J., *The Enigma War.* Scribner, New York City, 1979.

[9] Grätzer, G., *Math into Latex: An Introduction to Latex and AMS-Latex.* Birkhäuser, Boston, 1996.

[10] Gumm, H. P., A New Class of Check-Digit Methods for Arbitrary Number Systems. *IEEE Transactions on Information*, 31, 1985, 102–105.

[11] Gumm, H. P., Encoding of Numbers to Detect Typing Errors. *International Journal of Applied Engineering Education*, 2, 1986, 61–65.

[12] *The ISBN System Users' Manual.* International ISBN Agency, Berlin, 1986.

[13] Khan, D., *Seizing the Enigma: The Race to Break the German U-Boat Codes, 1939–1943.* Houghton Mifflin, Boston, 1991.

[14] Koblitz, N., *A Course in Number Theory and Cryptography.* Springer-Verlag, New York City, 1987.

[15] Lay, D. C., *Linear Algebra and Its Applications.* Addison-Wesley, Reading, Mass., 1994.

[16] Lewin, R., *The American Magic: Codes, Ciphers, and the Defeat of Japan.* Farrar Straus Giroux, New York City, 1982.

[17] Meier, J., and Rishell, T., *Writing in the Teaching and Learning of Mathematics.* The Mathematical Association of America, Washington, D.C., 1998.

[18] Prados, J., *Combined Fleet Decoded: The Secret History of American Intelligence and the Japanese Navy in World War II.* Random House, New York City, 1995.

[19] Rivest, R. L., Shamir, A., and Adleman, L., *A Method for Obtaining Digital Signatures and Public-Key Crypotosystems.* Laboratory for Computer Science, M.I.T., LCS/TM-82 (April 1977).

[20] Rivest, R. L., Shamir, A., and Adleman, L., A Method for Obtaining Digital Signatures and Public-Key Crypotosystems. *Communications of the ACM*, 21(2), 1978, 120–126.

[21] Schay, G., *Introduction to Linear Algebra.* Jones and Bartlett Publishers, Sudbury, Mass., 1997.

[22] Sethi, A. S., Rajaraman, V., and Kenjale. P. S., An Error Correcting Scheme for Alphanumeric Data. *Information Processing Letters*, 7, 1978, 72–77.

[23] Sinkov, A. *Elementary Cryptoanalysis: A Mathematical Approach.* Random House, New York City, 1968.

[24] Sterrett, A., ed., *Using Writing to Teach Mathematics.* MAA Notes no. 16, The Mathematical Association of America, Washington, D.C., 1990.

[25] Tuchinsky, P. M., International Standard Book Numbers. *UMAP Journal*, 5, 1989, 41–54.

[26] Verhoeff, J., *Error Detecting Decimal Codes.* Mathematical Centre, Amsterdam, 1969.

[27] Welchman, G., *The Hut Six Story: Breaking the Enigma Codes.* McGraw Hill, New York City, 1982.

[28] Welsh, D., *Codes and Cryptography.* Oxford University Press, New York City, 1988.

[29] Winters, S. J., Error Detecting Schemes Using Dihedral Groups. *UMAP Journal*, 11, 1990, 299–308.

INDEX

abelian, 143
associative, 126

bar graph, 101
border, 116

cardinality, 62
Cayley table, 137
characters, 2
check digit, 5
check digit scheme, 5
  airline ticket, 30
  German Bundesbank, 160
  IBM, 93
  International Standard Book Number (ISBN), 42
  Manitoba province driver's license, 69
  Universal Product Codes (UPC), 34
  United States postal money order, 26
  Verhoeff, 154
  Washington State driver's license, 69
cipher, 47
  Caesar, 47
  Enigma code 46
  monoalphabetic, 48
  plaintext shift, 47
  Purple code 46
  RSA public-key, 51
  substitution, 48
ciphertext, 46
collision, 67
composite, 12
composition of functions, 128
congruent, 23,
counting numbers, 9
country number, 41
cryptography, 46
cycle, 85

deciphering, 47
dihedral group, 109, 132
  D_{10}, 108, 132
disjoint, 87
divide, 12
Division Algorithm, 16
domain, 77
dot product, 35, 42

empty set, 62
enciphering, 47
encryption, 47

function, 74

group, 126

hashing, 66

identification numbers, 1
  airline ticket, 30
  German Bundesbank, 160
  International Standard Book Number (ISBN), 3, 6, 41
  length, 2
  Manitoba Province driver's license, 67
  Universal Product Codes (UPC), 2, 6, 34
  United States postal money order, 2, 6, 26
  Vehicle Identification Number (VIN), 3, 6
  Washington State driver's license, 3, 6, 67
identity, 126
integers, 9
intersection, 63
inverse, 126
irrational numbers, 10

mod n, 20, 21

0794